FPInnovations

Performance, Analysis, and Design of Mass Timber Diaphragms

Authors:
Marjan Popovski, Ph.D., P.Eng.
Samuel Cuerrier Auclair, M.Sc., P.Eng.
Zhiyong Chen, Ph.D., P.Eng.

Reviewer:
Chun Ni, Ph.D., P.Eng

Special Publication SP-546

2023
Pointe-Claire, Que.

Performance, Analysis, and Design of Mass Timber Diaphragms
(September 2023)

Special Publication SP-546

ISBN 978-0-86488-618-7 (print)
ISBN 978-0-86488-619-4 (digital)
ISSN 1925-0495 (print)
ISSN 1925-0509 (digital)

The information contained in this Work represents current research results and technical information made available from many sources, including researchers, manufacturers, and design professionals. The information has been reviewed by professionals in wood design including professors, design engineers, and wood product manufacturers. While every reasonable effort has been made to ensure the accuracy of the information presented, and special effort has been made to assure that the information reflects the state-of-the-art on the date of its going to press, neither FPInnovations, nor the above-mentioned parties make any warranty, expressed or implied, or assume any legal liability or responsibility for the use, application of, and/or reference to opinions, findings, conclusions, or recommendations included in this published Work, nor assume any responsibility for the accuracy or completeness of the information or its fitness for any particular purpose.

This published Work is not a substitute for professional advice. It is the responsibility of users to exercise professional knowledge and judgment in the use of the information.

Product or corporate names may be trademarks or registered trademarks, and are used only for identification and explanation without intent to infringe.

FOREWORD

Diaphragms are fundamental parts of the lateral load resisting system (LLRS) of any building, regardless of the material used. They not only transfer horizontal loads to the LLRS but also tie together all structural and non-structural elements, providing integrity to the entire building. The current worldwide trend to build with mass timber has led to the introduction of mass timber in many diaphragm applications. Mass timber diaphragms consist of larger and thicker timber panels made of products such as glued laminated timber (GLT or glulam), cross-laminated timber (CLT), nail-laminated timber (NLT), and dowel-laminated timber (DLT). Despite their fundamental roles in building performance, how mass timber diaphragms and their components perform under in-plane loads remains relatively under-researched. In addition, design guidelines for mass timber diaphragms are not as comprehensive as those for other parts of mass timber buildings in Canadian, U.S., and international building codes and material standards. Consequently, the main objective of this document is to present state-of-the-art information on the performance of mass timber diaphragms under in-plane loads, illuminate the methods used to analyze mass timber diaphragms, and summarize the available information on the design of mass timber diaphragms available in the Canadian, U.S., and international building codes and material standards.

ACKNOWLEDGEMENTS

This project was financially supported by Natural Resources Canada (NRCan).

INQUIRIES

Please submit inquiries about this document to:

FPInnovations
570 Saint-Jean Blvd
Pointe-Claire, QC
Canada H9R 3J9
web.fpinnovations.ca/
info@fpinnovations.ca

Follow us on:

Table of contents

1 INTRODUCTION 1

1.1 General 1

1.2 Main Diaphragm Components 2

2 TYPES OF MASS TIMBER DIAPHRAGMS 5

2.1 Glulam Decking 5

2.2 CLT Diaphragms 7

2.3 Nail-Laminated Diaphragms 10

2.4 Dowel-Laminated Diaphragms 11

2.5 Timber-Concrete Non-Composite and Composite Diaphragms 12

3 ANALYSIS OF MASS TIMBER DIAPHRAGMS 13

3.1 Diaphragm Loads 13

3.2 Method of Analysis 14

3.2.1 Deep Beam Analogy 15

3.2.2 Equivalent Truss Method 17

3.2.3 Finite Element Modelling and Analysis 23

3.3 Diaphragm Stiffness and Deflection 27

3.3.1 Basics of Simplified Diaphragm Deflection 27

3.3.2 Bending Deformation Contribution 28

3.3.3 Shear Deformation Contribution 30

3.3.4 Fastener Slip 31

3.3.5 Chord Spice Split 35

4 STIFFNESS AND SLIP OF MASS TIMBER CONNECTORS: CODES AND STANDARDS PERSPECTIVE 37

4.1 Connection Stiffness in Testing Standards 37

4.2 Determination of the Yield Deformation and Yield Load 39

4.3 Stiffness of Connections in Design Standards 41

4.3.1 Eurocode 5 41

4.3.2 Swiss Standard SIA 265 42

4.3.3 New Zealand/Australian Standard NZS AS 1720.1 43

4.3.4 Canadian Standard for Engineering Design in Wood CSA 086 44

5 PERFORMANCE OF CONNECTIONS IN MASS TIMBER DIAPHRAGMS 45

5.1 Half-Lap and Spline Connections Using STSs in CLT 45

5.2 Connections in CLT Using Double Inclined STSs 50

5.3 Connections in CLT with STSs in Shear and Withdrawals 52
5.4 Group Effect in CLT Connections Using STSs 54
5.5 Effect of the Angle of STSs on the Connection Properties 57
5.6 Half-Lap and Spline Connections Using STS in CLT Conducted at Oregon State University 60
5.7 Half-Lap and Spline Connections in CLT Conducted at Texas A&M University 66
6 EXPERIMENTAL AND ANALYTICAL PERFORMANCE OF MASS TIMBER DIAPHRAGMS 69
6.1 Tests on Diaphragms at FPInnovations 69
6.1.1 Configuration 1 (GL-89) 70
6.1.2 Configuration 2 (GL-89S) 71
6.1.3 Configuration 3 (GL-44) 71
6.1.4 Configuration 4 (GL-44S) 72
6.1.5 Configuration 5 (GL-44P) 72
6.1.6 Configuration 6 (CLT-105) 72
6.1.7 Configuration 7 (CLT-105SP) 73
6.1.8 Results and Discussion 73
6.2 Nordic Tests at FPInnovations 75
6.2.1 Results and Discussion 78
6.3 Research on Diaphragm Performance Conducted at UBC 79
6.4 Analytical Research on the In-Plane Flexibility of CLT Diaphragms 82
6.5 Testing of CLT Diaphragms in the U.S. 86
6.5.1 Testing at Colorado State University 86
6.5.2 American Wood Council Diaphragm Tests 89
6.5.3 Shaking Table Tests on Mass Timber Diaphragms 92
6.6 Testing and Numerical Studies on Hybrid Mass Timber Diaphragms 94
7 DESIGN OF MASS TIMBER DIAPHRAGMS 96
7.1 Diaphragm In-Plane Flexibility 97
7.2 Capacity-Based Design Principles 98
7.3 Design Principles in Canadian, U.S., and International Codes and Standards 99
7.3.1 The Canadian Standard for Engineering Design in Wood 99
7.3.2 Special Design Provisions for Wood and Seismic in the U.S. 100
7.3.3 European Codes and Standards 102
8 CONCLUDING REMARKS 105
9 REFERENCES 106

List of figures

Figure 1. Main diaphragm components. 3

Figure 2. Basic transfer of forces in a diaphragm, using deep beam analogy. 4

Figure 3. Irregular floor geometry diaphragms with typical components. 5

Figure 4. Fondaction CSN building and its structure. 6

Figure 5. Details of the connections from the glulam beams and decking to the concrete core walls 6

Figure 6. Two orthogonal directions for a floor diaphragm with glulam decking: (a) applied load is perpendicular to the decking panel length; (b) applied load is parallel to the decking panel length 7

Figure 7. CLT floor diaphragms used in hybrid construction: (a) Brock Commons building, (b) Tallwood 1 building. 8

Figure 8. CLT floor diaphragms used in all–mass timber construction: (a) Origine building, (b) Arbora building 8

Figure 9. Connections between adjacent CLT floor panels using (a) a surface spline, (b) a half-lap connection. 9

Figure 10. Half-lap joint adjacent to a side cantilever CLT panel 9

Figure 11. Nail-laminated timber 10

Figure 12. (a) Dowel-laminated timber (DLT); (b) use of DLT as a diaphragm in an industrial type of a building 11

Figure 13. (a) Example of timber-concrete composite (TCC) panel, (b) use of self-tapping screws (STSs) in construction of timber-concrete composite diaphragm. 12

Figure 14. (a) Concentrated loads introduced on diaphragm framing elements; (b) force transfer into diaphragm via fasteners in the sheathing panels for mass timber diaphragm. 14

Figure 15. Girder analogy for regular diaphragms. 16

Figure 16. Ritter's original truss model. 17

Figure 17. Shear panel with panel and fastener stiffnesses and equivalent truss diagonal 18

Figure 18. Idealization of wood-frame and mass timber diaphragm panel in a truss model with multiple diagonals. 21

Figure 19. Multiple diagonals for regular m x n subdivision (left) or for irregular subdivision (right). 22

Figure 20. Diaphragm FE model with discrete fasteners (springs) used in model panel splices and connections of the panels to the chords and collectors. 25

Figure 21. Different diaphragm models: (a) homogeneous model; (b) discrete model with 2-D connection zones; (c) discrete panel model with corner connections; and (d) discrete panel model with spaced connections 26

Figure 22. Idealized diaphragm represented as a deep I-beam. 27

Figure 23. Simplified diaphragm cross section used to determine bending displacement. 29
Figure 24. In-plane shear deformations of LWF diaphragms. 31
Figure 25. Fastener slip deflection contribution parallel to the applied load. 32
Figure 26. Deflection contribution from fastener slip perpendicular to the applied load. 33
Figure 27. Fastener slip due to bending deformation. 34
Figure 28. Deflection of a diaphragm due to chord splice slip. 36
Figure 29. a) Load-deflection curve of a typical connection tested using the EN 26891 standard and b) the idealized load-deformation procedure according to EN 26891 37
Figure 30. Definition of the yield and the ultimate point according to EN 12512 for two well-defined linear parts (left); and when the test curve does not cover well-defined linear parts (right) 39
Figure 31. Definition of the yield point according to SIA 265. 40
Figure 32. Definition of the yield point according to ASTM D5652 and D5764. 40
Figure 33. Definition of the yield point according to ASTM D5652 and D5764. 41
Figure 34. Test matrix for CLT connections with STSs that apply to diaphragm situations 46
Figure 35. Test results from six tested configurations of CLT connections with STSs 47
Figure 36. Typical hysteresis loops from cyclic tests along with the backbone curves and monotonic curves for different connection configurations: a) Config. #1; b) Config. #2; c) Config. #11; d) Config. #3; e) Config. #4; f) Config. #12 49
Figure 37. Screw assembly layout for test specimen, with photo of detail 50
Figure 38. Load-displacement curves from static tests (top); Connection properties from monotonic tests (bottom) 51
Figure 39. Typical cyclic behaviour obtained from the cyclic tests (left); Connection properties from cyclic tests (right) 52
Figure 40. Test specimen layout and testing series: (a) small specimens with one shear plane; (b) medium and large specimens with two shear planes; test series overview (right) 52
Figure 41. Screw layout: shear (left); withdrawal (centre); combined (right). Fastener 1: ASSY 3.0 ECOFAST Screw PT 8 mm × 90 mm; fastener 2: ASSY Plus VG Screw FT 8 mm × 140 mm 53
Figure 42. Summary of the averages of the test results, with COV in parenthesis 54
Figure 43. Reduction in strength for n_{eff}, according to different design standards and equations 55
Figure 44. Group effect on (a) monotonic capacity; (b) cyclic loading capacity; (c) monotonic stiffness; and (d) cyclic loading stiffness. 56
Figure 45. Specimen configuration for all test series 58
Figure 46. Typical load-deformation curves from test series S1 to S7 (left) and test series S8 to S13 59
Figure 47. Summary of the test results, with the percentages of variability 60
Figure 48. Test matrix for the various diaphragm connection configurations 61

Figure 49. Perspective view of the half-lap (left) and spline (right) connection specimens ... 62
Figure 50. Results summary ... 62
Figure 51. Plan and cross-sectional view of the tested connections ... 64
Figure 52. Test matrix for spline and lap connection tests ... 64
Figure 53. Plan and cross-sectional views of the tested connections ... 65
Figure 54. Typical hysteretic curve obtained during testing of connections with STSs (left) and the connection strength and stiffness parameters per fastener ... 65
Figure 55. Test matrix CLT connections ... 67
Figure 56. Peak load, reference displacement, and peak load per fastener from the monotonic tests ... 67
Figure 57. Average force-displacement curve per fastener for the tested CLT shear connections ... 68
Figure 58. Test setup for determining the in-plane load-deformation behaviour of mass timber diaphragms. ... 70
Figure 59. Diaphragm configuration 1: 89 mm–thick glulam panels continuous along the diaphragm's length ... 70
Figure 60. Diaphragm configuration 2: 89 mm–thick glulam panels arranged in a staggered pattern. ... 71
Figure 61. Diaphragm configuration 3: 44 mm–thick glulam panels continuous throughout the diaphragm's length ... 71
Figure 62. Diaphragm configuration 4: 44 mm–thick glulam panels arranged in a staggered pattern. ... 72
Figure 63. Diaphragm configuration 5: 44 mm–thick glulam panels placed perpendicular to the diaphragm length. ... 72
Figure 64. Diaphragm configuration 6: 105 mm–thick 3-ply CLT panel. ... 73
Figure 65. Diaphragm configuration 7: 105 mm–thick 3-ply CLT panels fastened side-by-side. ... 73
Figure 66. Load-deflection curves for the seven tested diaphragms. ... 74
Figure 67. The test setup used for the Nordic diaphragms. ... 75
Figure 68. Diaphragm dimension with the position of the installed measuring devices. ... 76
Figure 69. OSB panel fixed to the glulam decking for configurations 1 and 2. ... 76
Figure 70. The nailed glulam decking of diaphragm configuration 1. ... 76
Figure 71. The nailed glulam decking of diaphragm configuration 3. ... 77
Figure 72. The loading protocol used to test all diaphragms. ... 77
Figure 73. Load–mid-span deflection curves for the 3 diaphragms. ... 78
Figure 74. Load–Relative slip between the glulam decking near the supports, for each diaphragm configuration. ... 79
Figure 75. Diaphragm configuration ... 80

Figure 76. In-plane stiffness of a CLT diaphragm for different initial stiffnesses of panel-to-panel connections ... 81

Figure 77. In-plane stiffness of a CLT diaphragm with different numbers of panels and two different in-plane shear moduli for the panels ... 82

Figure 78. An example of the plane model ... 83

Figure 79. Mechanical properties of connections in the plane model ... 83

Figure 80. Spring model for the in-plane behaviour of CLT diaphragms ... 84

Figure 81. Shear and global bending deformation contributions [in %] for different diaphragm widths B [in m] ... 85

Figure 82. A plan of the test setup ... 87

Figure 83. Connector layouts for configuration 1 with a weak tension chord connection (left) and configuration 2 with a stronger tension chord connection (right) ... 88

Figure 84. The result summary of the four tests ... 88

Figure 85. Load-displacement hysteretic curves from the four tests; configuration 1 (top and bottom left) and configuration 2 (bottom right) ... 89

Figure 86. Diaphragm assemblies for diaphragm CLT 01, with the long dimension of the CLT panels oriented parallel to the load (left) and CLT 02, where they are perpendicular to the load (right) ... 90

Figure 87. Load-deformation curves for both diaphragms: (a) average reaction versus centreline deflection; and (b) connection deformations ... 91

Figure 88. Two-story mass timber structure tested during the summer of 2017 at the University of California San Diego outdoor shaking table ... 92

Figure 89. Structural layouts for phase 1 of testing: floor level (top); roof level (bottom) ... 93

Figure 90. Hybrid floor system: (a) individual segment; (b) floor assembly; (c) beam-to-beam connections; (d) manufacturing process for the floor ... 95

Figure 91. Deflection in the middle of the diaphragms, and that of the LLRS underneath ... 97

Figure 92. Capacity design and chain analogy ($E_{seismic}$ is the seismic demand, $R_{n,\,ductile}$ is the strength of the ductile element, ϕ_o is the overstrength factor of the ductile element, R_{n1},and R_{n2} are strengths of the brittle element) ... 99

Figure 93. Typical storey of a multi-storey CLT structure with various connections between panels. Diaphragm connections 2 and 3 have to be elastic, while wall-to-wall connections 4 must be ductile. ... 100

Figure 94. The test setup, with diaphragm specimen no. 2 ready for testing ... 101

Figure 95. Openings in CLT diaphragms ... 103

Figure 96. Location of horizontal joints in CLT diaphragms vs. vertical joints in CLT walls ... 104

List of tables

Table 1. Available analysis methods for timber diaphragms and their requirements 15

Table 2. Values of the c_i coefficients for most used connections between diaphragm panels 20

Table 3. Effective area and elasticity for regular and irregular truss diagonal models 21

Table 4. Stiffness and strength properties of the tested diaphragms 74

Table 5. Load, deflection, and stiffness values from the diaphragm tests 79

1 INTRODUCTION

1.1 General

Timber structures provide a sustainable, energy efficient, durable, and relatively affordable solution to the growing demand for multi-storey residential and commercial buildings. This new interest in mid- and high-rise timber buildings has created a need for a more rigorous structural analysis and improved design approaches for all components of the structural system. This is especially true for buildings located in high seismic or wind areas that are subjected to high lateral loads. These lateral loads are resisted by the building's lateral load resisting system (LLRS), also known as the seismic force resisting system (SFRS) when the building is located in seismic prone areas and the earthquake loads govern the lateral load design.

Diaphragms play a fundamental role in the structural system of any building, regardless of the building material. They not only transfer horizontal loads to the LLRS but also tie all structural and non-structural elements together, providing integrity to the entire building. Diaphragms are a crucial component in a building, and the loss of diaphragm action could lead to the collapse of the building due to instabilities in vertical elements or the changed load distribution in the LLRS (Moroder, 2016). Their importance in structural performance, however, has been partially neglected in the past because most floors also possess some intrinsic in-plane strength and stiffness, which allow for some diaphragm action. This has often led to the misconception that all diaphragms are rigid and strong and therefore can withstand arbitrary loads and deformations without damage. This fallacy is partially attributable to shortcomings in design codes, the literature, and designers' education, which often neglect the design of diaphragms (Moroder, 2016).

The main role of the floor and roof diaphragms is to carry the gravity loads at each storey of the structural system and transfer it to the vertical elements of the building's gravity load resting system (GLRS). However, they play an equal, if not more important, role in case of lateral loads; as summarized below (Moehle et al., 2010). The diaphragms:

- **Transfer lateral force to the LLRS underneath** – The floor system usually contains most of the mass of the building at each storey. Consequently, significant inertial forces develop in the plane of the diaphragm during an earthquake. One of the primary roles of the diaphragms at each storey is to take these earthquake-induced forces from the floor system, as well as those from the tributary portions of walls and columns, and transfer them to the parts of the LLRS below. Additionally, diaphragms provide out-of-plane resistance to the exterior walls when subjected to wind loads and transfer those loads to the elements of the LLRS below the diaphragm.
- **Provide lateral stiffness to vertical elements** – Diaphragms are connected to the vertical elements of the LLRS at each storey, and they provide lateral support to help resist the buckling of the vertical elements of the gravity load resisting system (GLRS) that mainly consists of frames, or walls. They also help resist second-order forces associated with the axial forces acting through lateral displacements of the vertical elements of the LLRS. Furthermore, by tying together the vertical elements of the LLRS, the diaphragms complete a three-dimensional framework that can resist lateral loads.
- **Resist wall and façade out-of-plane forces** – As a building responds to an earthquake, the exterior walls, façade, and cladding develop out-of-plane lateral inertial forces. Wind pressure acting on exposed wall surfaces also creates out-of-plane forces, to which the diaphragm-to-wall connections provide resistance.

- **Transfer forces through discontinuities in the vertical LLRS** – As a building responds to an earthquake loading, it often becomes necessary to transfer lateral shear forces from one vertical element of the LLRS to another. The largest transfers usually occur when there are discontinuities in the vertical elements of the LLRS, especially in cases with in-plane and out-of-plane offsets (discontinuities). Large diaphragm transfer forces can occur in such cases, and the diaphragm elements need to be able to take and transfer them.
- **Resist thrust from inclined columns** – In cases where architectural configurations require inclined columns, gravity and overturning actions can generate large horizontal forces acting within the plane of the diaphragms. These forces can act either in tension or in compression, depending on the orientation and the stresses in the column, and the diaphragm and its components need to be designed to resist these forces.

Despite their fundamental role in building performance, how mass timber diaphragms and their components perform under in-plane loads remains under-studied. In addition, design guidelines for mass timber diaphragms are not as comprehensive as those for other parts of mass timber buildings in the Canadian, U.S., and international codes and material standards. Consequently, the main objectives of this document are to:

a) Present state-of-the-art experimental and analytical research on the performance of mass timber diaphragms around the world;

b) Provide up-to-date information on methods of analysis for mass timber diaphragms;

c) Summarize the available information on the design of mass timber diaphragms available in the Canadian, U.S., and international codes and standards.

The current knowledge on this subject is not as comprehensive as for many other building components, and the information presented in this document may thus be updated in the future as more research and design knowledge becomes available.

1.2 Main Diaphragm Components

Diaphragms can be made of timber, reinforced or precast concrete, or a wide variety of steel. They can be made from the same material as the LLRS or from a different one, in case of hybrid buildings. In the latter case, the vertical LLRS can consist of reinforced concrete, steel, or masonry, while the diaphragms can vary between timber, steel, and reinforced concrete. The diaphragms themselves can also be made either of a single material or of composite materials, such as timber-concrete composite (TCC), timber or concrete composite steel decks, etc.

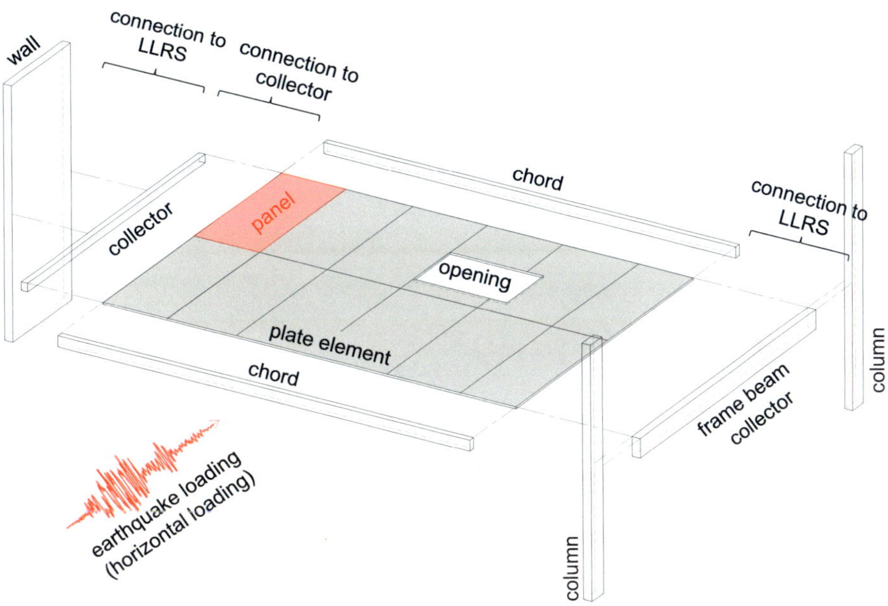

Figure 1. Main diaphragm components (modified from Moroder, 2016).

Regardless of the material used, diaphragms have similar main parts and components. The main components of any type of diaphragm appear in Figure 1 (Moroder, 2016) and can be grouped as follows:

- Plate element;
- Chords or chord beams;
- Collectors;
- Drag struts or drag beams;
- Connections to the LLRS underneath.

In general, diaphragm behaviour can be explained using a deep I-beam (girder) analogy, where the web is made of the "plate element" and the flanges consist of the chord beams (Figure 2). The "plate element" takes the shear forces (V_u) in such a deep beam configuration, while the cords take the bending actions (M_u) through the tension (T_u) and compression (C_u) forces in them. The "plate element" can consist of a single panel (as in cast in-situ reinforced concrete diaphragms), or it can be made of multiple panels with connections, as in the case of mass timber diaphragms. The collector beams take the force actions from the diaphragms through the connections and transfer them to the vertical LLRS.

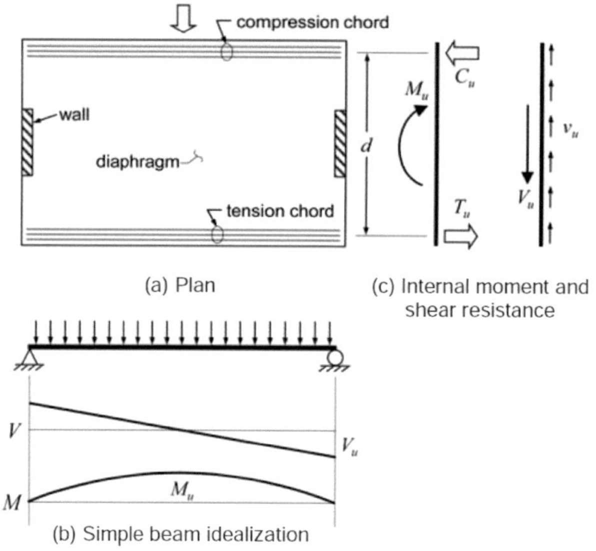

Figure 2. Basic transfer of forces in a diaphragm, using deep beam analogy (Moehle et al., 2010).

The diaphragms can be defined to be regular or irregular, with or without openings. Regular diaphragms usually have a rectangular shape, without any significant disruptions in the flow of forces within them. Diaphragms that have non-rectangular shapes, have discontinuous load paths, contain discontinuities, or miss elements such as chords, collectors, or drag struts, either in some places or overall, are usually called irregular diaphragms (Figure 3).

Drag struts or drag beams collect the shear forces from the disturbed areas around openings and corners and anchor them into adjacent parts of a diaphragm. These parts are commonly referred to as sub-diaphragms or transfer diaphragms (Diekmann, 1992; Malone & Rice, 2012). These terms should not be confused with diaphragms that resist transfer forces from displacement incompatibilities. One can often adopt reduced fastener spacing, increased reinforcement, or thicker framing members around irregularities, but sub-diaphragms are essentially designed as regular diaphragms. The resultant shear forces in the diaphragm have to be collected and conveyed to the LLRS via the collectors (i.e., collector regions or beams). The connections of the collector to the LLRS require proper design, as they are an essential part of the load path into the foundations. For mass timber diaphragms, this means that the panels need to be connected together and the beams acting as chords need to be connected to the plate to guarantee the force transfer.

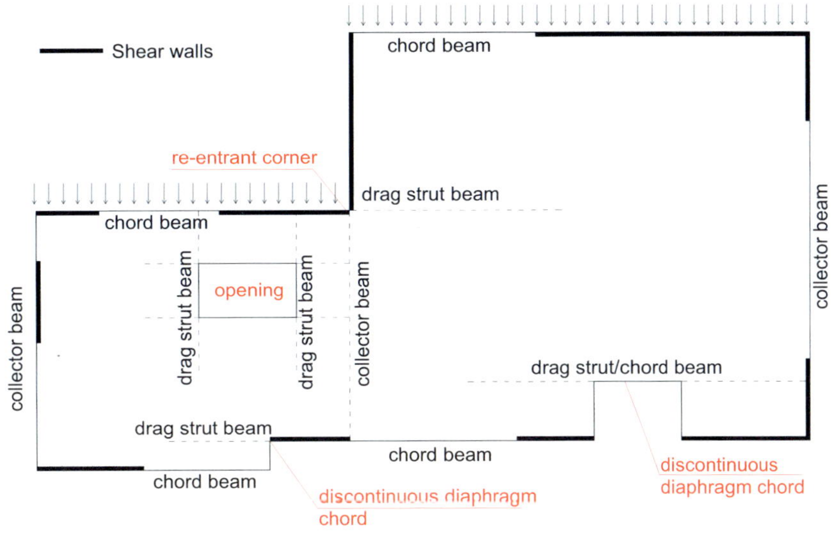

Figure 3. Irregular floor geometry diaphragms with typical components (modified from Moroder, 2016).

2 TYPES OF MASS TIMBER DIAPHRAGMS

The current worldwide trend to build with mass timber has led to the introduction of mass timber diaphragms. In comparison to light wood-frame (LWF) diaphragms that consist of lumber or I-joists sheathed with 4 x 8 ft. plywood or oriented strand board (OSB) panels on the top, mass timber diaphragms consist of larger and thicker timber panels and beams. Products used for the panels in mass timber diaphragms include glued laminated timber (GLT or glulam), cross-laminated Timber (CLT), nail-laminated timber (NLT), and dowel-laminated timber (DLT).

When analysing such diaphragms as a deep I-beam, the thick mass timber panels and beams take the applied bending moment and shear force. To ensure proper diaphragm actions, panels must connect to each other with appropriate connections and connection details, such as those discussed in Karacabeyli and Gagnon (2019). In general, the strength and stiffness of a diaphragm are strongly influenced not only by its material but also by the properties of the connections used. The increased in-plane strength and stiffness of the larger mass timber panels and the wide variety of new fasteners on the market has opened the possibility of using mass timber diaphragms in larger and taller timber buildings.

2.1 Glulam Decking

Glulam elements are often used as beam or column elements, but they can also function as decking elements that form the diaphragm. The Fondaction CSN building in Québec is a good example of using glulam in diaphragms. The floor consists of a thick glulam deck supported by glulam beams. These beams are 191 mm wide and 530 mm deep, spaced at 1.52 m o.c., with a clear span of 8.65 m. The glulam deck is 89 mm thick and 610 mm wide. A 15.5 mm–thick plywood sheathing is fastened to the glulam deck to increase the in-plane stiffness and strength. Individual glulam deck panels connect to each other with OSB splines. The LLRS of the building is made of concrete shear walls and cores, as shown in Figure 4. Figure 5 shows the connection between the floor diaphragm and the LLRS.

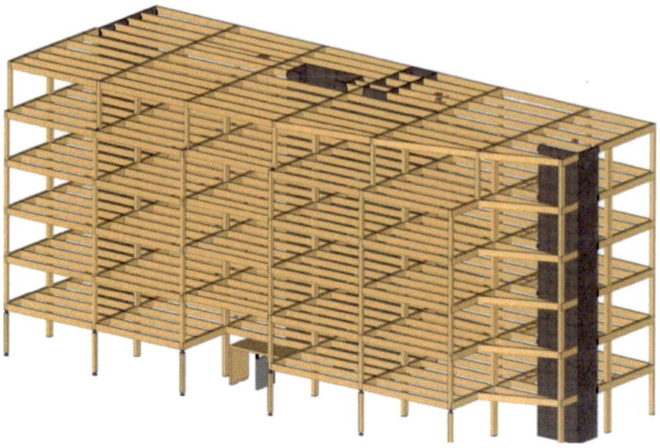

Figure 4. Fondaction CSN building and its structure (courtesy of Nordic Structures).

Figure 5. Details of the connections from the glulam beams and decking to the concrete core walls (courtesy of Nordic Structures).

This type of mass timber diaphragm behaves differently depending on the direction of the glulam decking with respect to that of the load. When the loading is perpendicular to the length of the glulam, in addition to the chords and the connections, the decking itself can also resist some of the applied bending moment (Figure 6a). In the other direction, the decking itself is less involved in resisting the bending moment, and its resistance comes more from the connections between the decking and the beams underneath, as well as from the connections between the decking panels, if present (Figure 6b). Consequently, the diaphragm is much more flexible when the load direction is the same as that of the decking panels.

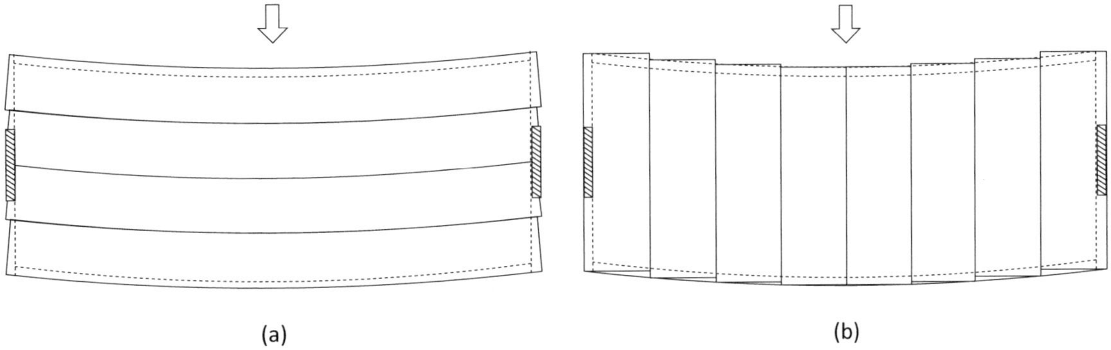

Figure 6. Two orthogonal directions for a floor diaphragm with glulam decking: (a) applied load is perpendicular to the decking panel length; (b) applied load is parallel to the decking panel length.

Glulam decking design should comply with clause 7 on glulam in CSA O86-19 (CSA, 2019), or more specifically, with Clause 7.5.3 on vertically glued laminated beams. If the properties of the glulam decking differ from those provided in O86, the appropriate Canadian Construction Materials Centre report for the product can provide the properties.

2.2 CLT Diaphragms

Cross-laminated timber (CLT) is one of the most common mass timber panels; thus, it is common to see it used in the floor and roof diaphragms of mass timber buildings. Panel sizes vary by manufacturer; typical widths range from 0.6 to 3 m (2.0 to 10 ft.), the length can be up to 18 m (60 ft.), and the thickness can go up to 508 mm (20 in.). Lately, panel widths up to 5.5 m (18 ft.) and lengths up to 29.9 m (98 ft.) have become possible, though uncommon. In some cases, the size of the panels for a project can be limited by transportation constraints. Hundreds of mass timber buildings with variable heights built around the world in the past 20 years have used CLT in wall and/or diaphragm applications. CLT diaphragms are options in all mass timber buildings or in hybrid buildings where the vertical LLRS is of either steel or concrete. Examples of CLT diaphragms used in hybrid buildings in Canada include the Brock Commons building (Figure 7a) and the Tallwood 1 building in Langford, B.C. (Figure 7b). The LLRS in the 18-storey Brock Commons building includes concrete cores and CLT diaphragms (Figure 7a), while the 12-storey Tallwood 1 building uses steel-braced frames as the main vertical LLRS, along with CLT diaphragms.

(a) (b)

Figure 7. CLT floor diaphragms used in hybrid construction: (a) Brock Commons building, (b) Tallwood 1 building.

(a) (b)

Figure 8. CLT floor diaphragms used in all–mass timber construction: (a) Origine building, (b) Arbora building (photos courtesy of Nordic Structures).

Examples of CLT floor diaphragms used in all–mass timber construction include the 13-storey Origine building in Québec City (Figure 8a) and the Arbora building complex in Montréal (Figure 8b). The CLT panels of the diaphragms can be supported by beams or walls, in which case the panels are analysed and designed to resist out-of-plane gravity loads mostly as unidirectional slabs. When the diaphragm CLT panels are supported on columns, the design can take advantage of the two-way spanning capability of CLT panels, as is the case with the Brock Commons and Tallwood 1 buildings.

In comparison to glulam decking, CLT panels have good in-plane stiffness and strength in both orthogonal directions due to their cross-lamination. Since the panels are significantly longer than they are wide, the CLT diaphragms may have different stiffness and strength in the two in-plane orthogonal directions. This further depends on the type and number of connections used between the diaphragms and the members of the LLRS, as well as on the connections between the CLT panels. If one wants a stiff and strong diaphragm effect, the connection between the panels needs to be designed accordingly. Typically, adjacent CLT panels are connected to each other with a surface spline or a half-lap joint (Figure 9).

(a)

(b)

Figure 9. Connections between adjacent CLT floor panels using (a) a surface spline, (b) a half-lap connection (courtesy of WoodWorks).

In spline joints, a shallow rabbet is routed in the top of each adjacent panel, and a narrow plywood or laminated veneer lumber (LVL) strip is nailed across the joint in the field. Nails are typically used when a more flexible connection is necessary and relatively low many pull-out forces are expected. For stiffer connections, or when some fastener pullout is expected, self-tapping screws (STS) are an option. Half-lap joints between adjacent CLT panels use STSs in almost all cases. For floor applications, spline joints are often preferrable to the half-lap joints. The latter are more costly to fabricate than a surface spline joint, and they slightly reduce the effective width of the panels, thus increasing the total amount of wood a project requires. Half-lap joints would be appropriate when a CLT panel connects to an adjacent side cantilever panel (Figure 10) and the joint must resist some uplift. STSs driven at angles between 30 and 60 degrees to a panel offer higher strength and stiffness but lower flexibility in both types of connections.

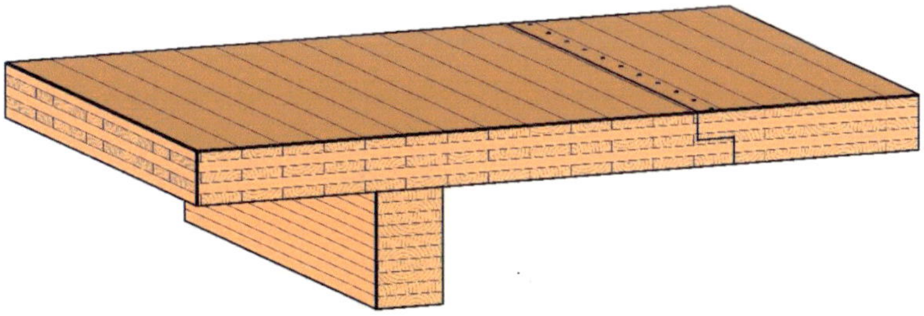

Figure 10. Half-lap joint adjacent to a side cantilever CLT panel (DeStefano, 2019).

In case of CLT diaphragms, the diaphragm boundary elements (chords and collectors) often consist of glulam or other mass timber beams. Connections at the ends of the timbers, which serve as boundary elements, must be detailed to resist the axial forces induced into them and thus maintain the continuity of the chords. If it is the CLT panels rather than the timber beams that must resist the chord forces, steel straps are necessary at the panel joints. To preclude non-ductile failure in the cords, any chord splices should be designed with sufficient overstrength (DeStefano, 2019). The STSs fastening the CLT panels to the boundary timbers need to be designed and spaced properly to transmit the chord and collector forces. Different connection details for CLT panels and timber elements appear in Chapter 5 of the FPInnovations CLT Handbook (Karacabeyli and Gagnon, 2019).

2.3 Nail-Laminated Diaphragms

Nail-laminated timber (NLT) is an almost 150-year-old method of timber construction that has seen a recent renaissance. Engineers and architects have rediscovered NLT as a new addition to mass timber products, as it shares the many benefits of the mass timber family in terms of wood technology and manufacturing. NLT is created by placing dimension lumber, of 38, 64, or 89 mm (nominal 2, 3, or 4 in.) width, and 89 to 286 mm (nominal 4 to 12 in.) depth, on edge, and fastening the individual laminations together with nails (Binational Softwood Lumber Council and Forest Innovation Investment [BSLC and FII], 2017). Typically, NLT (Figure 11) is used for floors and roofs, but in some instances, the panels can also be used for walls, elevator shafts, and stair shafts.

Figure 11. Nail-laminated timber (photo courtesy of StructureCraft).

NLT has good bending properties for out-of-plane loading. In-plane load transfer across lamination joints, however, is not well understood, nor is the contribution of those joints to NLT's in-plane shear and bending stiffness. The in-plane properties of NLT panels also differ in both perpendicular directions. For these reasons, plywood or OSB sheathing nailed to one face of the NLT helps provide in-plane shear capacity for diaphragm action. As a conservative approach, design values for NLT diaphragms sheathed with plywood or OSB should be taken from the diaphragm capacities given in CSA O86 (CSA, 2019) for blocked plywood/OSB diaphragms. It is also possible to estimate other diaphragm properties, such as deflection and flexibility, using the principles in CSA O86-19 for fully blocked diaphragms.

For diaphragms, the line of nails-to-plywood/OSB panel joints parallel to the direction of the NLT span should always be centred on an individual lamination to allow for proper load transfer across the joint. Where multiple rows of fasteners are necessary at panel edges and boundaries, the NLT laminations must have sufficient nailing between them to transfer the shear load across the joint. A simple approach is to provide equal nailing between the laminations and the plywood/OSB panel joint. Another common approach is to provide STS reinforcement at the NLT edges near the plywood/OSB splices (BSLC and FII, 2017).

The tension and compression forces in the chords and collectors in NLT diaphragms are usually resisted using beams as axial force members. When beams are not included at the edges of the diaphragm, chord forces must be resisted within the NLT floor assembly. One approach is to assume that one or several lamination(s) act as discrete tension and compression elements that resist the chord forces at the edges of the diaphragm. The lamination(s) should therefore be designed to carry combined axial and bending loads, with the axial forces due to lateral loads and the bending forces due to gravity loads in each load combination. If more than one lamination serves as a cord member, the laminations must be sufficiently nailed together for proper shear transfer. Compression force transfer across the lamination butt joints can be provided by direct end-grain bearing. One can transfer tension across a butt joint by transferring the force into the adjacent lamination and then back into the original lamination on the other side of the joint, using nails in shear. However, this load path is complicated when multiple laminations are necessary to resist the tension force and for layups with frequent butt joints between supports. The use of light-gauge steel straps provides a simpler approach.

2.4 Dowel-Laminated Diaphragms

Dowel-laminated timber (DLT, Figure 12a) is a mass timber panel created by placing dimension lumber, of 38, 64, or 89 mm (nominal 2, 3, or 4) width, and 89 to 286 mm (nominal 4 to 12 in.) depth, on edge, then fastening the individual laminations together with hardwood dowels. In many instances, DLT panels are similar to NLT panels, and they can be used mostly in floor and roof applications (Figure 12b).

(a) (b)

Figure 12. (a) Dowel-laminated timber (DLT); (b) use of DLT as a diaphragm in an industrial type of a building (photos courtesy of StructureCraft).

To form DLT members, softwood lumber panels are stacked like NLT and friction fits them together with hardwood dowels. The dowels hold the boards together, and the friction fit is achieved by the different moisture content of the softwood boards and the hardwood dowels, which provides dimensional stability to the entire panel. In general, dowels that hold each board side by side form a stiffer and stronger connection than the nails in NLT. Typically, each board lamination in a DLT panel is also finger-jointed. This creates a stiffer and stronger panel than NLT, as it eliminates the board splices and butt joints that are characteristic of NLT; admittedly, however, NLT can also use finger-jointed lamination. Dowels can also be inserted diagonally in which case they can provide additional resistance and stiffness to the entire panel.

DLT panels are prefabricated in sizes of up to 3.6 m (12 ft.) wide and 18.3 m (60 ft.) long. Each panel is put through a panel planer to ensure a dimensionally accurate and planed surface. Prefabricated panels can be factory finished with sealers or stains. Similar to NLT or glulam decking on flat, all the wood fibre runs in the direction of the span. This provides efficient use of material for floor and roof systems. Plywood or OSB sheathing needs to be nailed on top of the DLT panel to provide in-plane shear capacity, which one can calculate the same as a typical nailed plywood diaphragm, as per CSA O86.

2.5 Timber-Concrete Non-Composite and Composite Diaphragms

In general, most mass timber floor panels have non-structural concrete topping on the top to satisfy the code requirements for acoustic and fire performance. This concrete topping may or may not contain steel reinforcement, and it is poured on top of the panel directly or through an elastic and resilient underlayment. Therefore, there is no composite action between the unbounded concrete layer and the wood floor. Structural calculations mostly consider this topping as a dead weight and therefore do not consider it as contributing to the in-plane strength and stiffness of the diaphragm. Although no study has quantified the effect of this topping on the in-plane stiffness of the diaphragm, the topping may increase diaphragm stiffness, especially for resisting wind loads. The concrete layer on top of the wood floors adds additional mass to the floor, which reduces the floor vibration performance; it also adds stiffness to the floor, which improves the floor vibration performance. Consequently, the unbonded concrete layer thickness must not only meet the fire requirements but also not reduce the floor vibration performance.

(a) (b)

Figure 13. (a) Example of timber-concrete composite (TCC) panel, (b) use of self-tapping screws (STSs) in construction of timber-concrete composite diaphragm (photos courtesy of StructureCraft).

For the diaphragm to act as timber-concrete composite (TCC) both in-plane and out-of-plane, the concrete topping has to be structural and must be connected via steel shear transfer connections to the timber panel (Figure 13a). Shear transfer connections can vary and may include various types of perforated steel plates, or various types of STS. The CLT design handbook (Karacabeyli & Gagnon, 2019) presents several types of connections for CLT products. The use of TCC optimizes the performance and material requirements and provides structural efficiency by creating composite action between the two materials. This in turn increases the available spans, reduces deflections, and improves vibration performance, thus providing a more cost-efficient solution for the floor and diaphragms. TCC also makes the diaphragm much stiffer than with mass timber panels only. Furthermore, the timber members can transfer tension and compression forces and therefore act as collector or chord beams.

The added concrete in both non-composite and composite floors should be accounted for as dead weight when analysing the structure under earthquakes or high winds.

3 ANALYSIS OF MASS TIMBER DIAPHRAGMS

3.1 Diaphragm Loads

Diaphragms take loads (actions) generated from seismic or wind actions, from transferred forces in the diaphragm, concentrated forces from inclined columns or from the restraint of vertical elements. These actions can be further differentiated by the type of load application: (a) area (surface) loads, (b) line loads, and (c) discrete loads. Seismic loads are typical representatives of area loads, while wind loads generate a line load on the diaphragms. Discrete or concentrated loads can be a result of vertically offset walls. Transfer forces can be treated like imposed displacements from the LLRS and have the same effects as concentrated loads. Most building codes and designers treat the diaphragm as a simply supported (or continuous) beam loaded with a uniformly distributed load. Usually, they do not distinguish whether the loads are applied to the compression or tension edge, applied to the panels or framing elements, or distributed over the diaphragm panels. A better approach for seismic loads, simply as an example, would be to distribute the inertia loads according to the seismic masses, which is how these loads are generated in reality.

A comparative study (Moroder et al., 2015) showed that in wood-frame diaphragms, the type of load application does not influence the load path, as long as the framing elements serve to both introduce and distribute forces. In mass timber diaphragms, force introductions along diaphragm edges generate longitudinal stresses in the diaphragm panels and create force components perpendicular to the panel edges, to be resisted by the connections. It is also necessary to transfer longitudinal stresses along the panels to activate diaphragm portions away from the point of force introduction.

Wind actions are based on the normal pressure created by blowing wind, in conjunction with the shape and orientation of the building, which can create increases in pressure and sometimes dynamic effects that require design attention. For large wind loads and larger diaphragm spans, horizontal loads on façade elements can become substantial and one must take care that the horizontal forces are transferred through the diaphragm into the LLRS. Loads applied to a diaphragm edge, from a combination of wind suction and internal pressure, for example, may not necessarily activate the entire diaphragm, and there must be load paths between framing to ensure that the force components perpendicular to the panel edges do not pull the diaphragm elements apart.

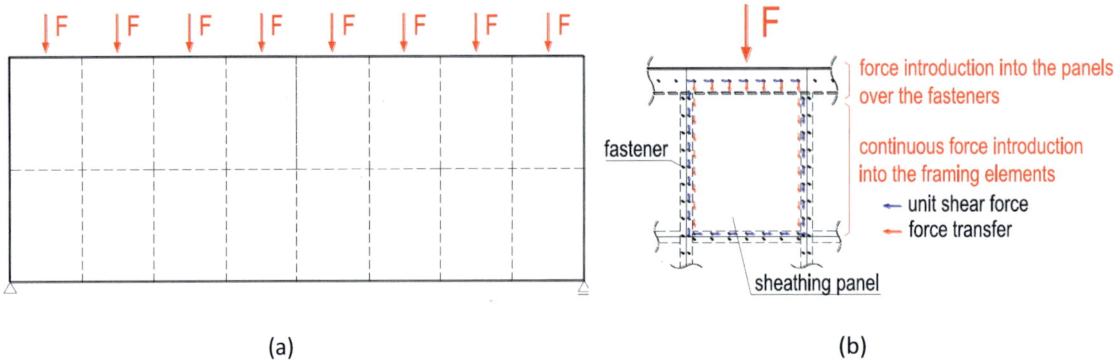

Figure Error! No text of specified style in document.**14. (a) Concentrated loads introduced on diaphragm framing elements; (b) force transfer into diaphragm via fasteners in the sheathing panels for mass timber diaphragm (Moroder 2016).**

The wind load acting on the diaphragm edge (Figure **Error! No text of specified style in document.**14a) must be transferred into the body of mass timber diaphragm. The forces from the chord beam are transferred into the timber sheathing panels, which requires the fasteners to resist forces perpendicular to the panel edges. Moreover, this mechanism also creates normal panel stresses. Since the fasteners already need to resist the unit shear force (the blue arrows in Figure **Error! No text of specified style in document.**14b) from the diaphragm action, it is necessary to check if they can also resist the additional load component (red arrows) from the direct load introduction. The additional force component normally also requires increased edge distance for the fasteners. A simple approach to take the resultant force, which accounts for both the unit shear force and the additional transfer force component, is to reduce the diaphragm depth and therefore increase the unit shear force in the panel fasteners.

One can determine seismic loads on a building based on its period (stiffness), the seismic weight, the seismic hazard of the location, and the anticipated building response parameters. The obtained seismic loads are then applied at each storey of the building. Although this concept is clear, its interpretation differs around the world, and the calculation of the diaphragm seismic demand in timber structures varies. In addition, comprehensive code provisions for the seismic analysis and capacity design of timber structures and their diaphragms are still poor or absent.

3.2 Method of Analysis

Several different analysis methods and approaches for diaphragms are available in the literature. Table 1 provides a quick overview of available analysis methods applicable to timber diaphragms, along with the requirements needed and the features they can account for. Designers need to decide which approach best suits a given problem and which level of accuracy they require.

Table 1. Available analysis methods for timber diaphragms and their requirements

Analysis method	Continuous cords	Unsupported edges	Concentrated loads	Openings/re-entrant corners	Deformations
Deep beam analogy	Required	Usually no allowed*	Allowed	Not allowed	Determined using equation for uniformly distributed loads
Shear field analogy	Required	Allowed under certain circumstances	Not allowed	Allowed	Not determined
Truss analogy	Required	Allowed under certain circumstances	Allowed	Allowed	Determined
Finite element analysis	Not required, as timber panels can transfer chord forces	Allowed	Allowed	Allowed	Determined

* Some design standards provide guidance or tabulated reduced capacities

3.2.1 Deep Beam Analogy

The simplest approach, and one of the most common methods to analyse diaphragms is the deep beam analogy method. This method is especially efficient and accurate for regular and rectangular diaphragms. It determines the load path and the internal actions in the diaphragm components by assuming the entire diaphragm as on deep beam or girder (Figure 15). In such case, the diaphragm panels resist the shear forces that are assumed to be constant along the diaphragm depth, while the diaphragm chord beams, acting in tension and compression, resist the bending moment.

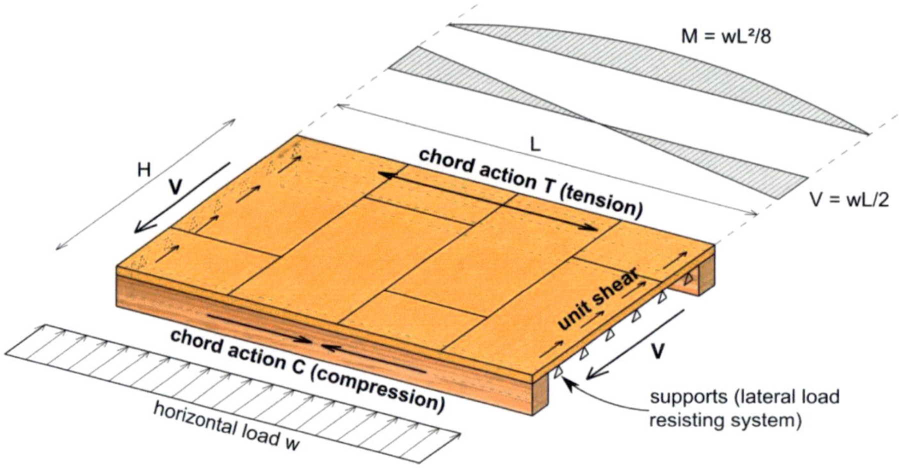

Figure 15. Girder analogy for regular diaphragms (Moroder 2016).

$$T = C = \frac{M}{H} = \frac{wL^2}{8H} \tag{1}$$

where

- T is the tension force in the chord;
- C is the compression force in the chord;
- w is the uniformly distributed load;
- L is the diaphragm span;
- H is the diaphragm depth; and
- M is the bending moment.

The unit shear force, defined as the shear force per unit length (or shear flow), can be calculated as follows:

$$v = \frac{V}{H} = \frac{wL}{2H} \tag{2}$$

where

- v is the unit shear force; and
- V is the shear force at diaphragm supports.

For diaphragms with few irregularities, hand methods based on first principles are still feasible to determine unit shear forces and axial forces in framing elements. Please refer to Malone and Rice (2012) for the analysis of irregular LWF diaphragm based on first principles.

3.2.2 Equivalent Truss Method

While the deep beam analogy method can be successfully used for design of light-frame diaphragms (Smith et al. 1986) and studies have shown it to be adequate for mass timber diaphragms (Wallner-Novak et al., 2014; Moroder et al., 2014, 2015), it is not sufficiently accurate to design modern floors with irregular geometries. Diaphragm openings, re-entrant corners, setbacks, and other irregularities create shear force concentrations, force components perpendicular to the panel edge, and decreased diaphragm stiffness. The deep beam analogy method cannot account for all these effects.

The truss analogy is an old concept, first developed in 1899 by Ritter for the analysis of reinforced concrete components (Kim, 2021). It describes the flow of forces in a reinforced concrete component after a concrete member cracks by using a truss model, as shown in Figure 16, for a deep, simply supported beam. The assumption is that the flow of force within a reinforced concrete component is the same as that within a truss model. This truss model comprises compressive stress fields connected via tension ties. The compression zone consists of the upper chord, which is mainly made of concrete that bears the compressive stress acting on the member, along with the inclined diagonal compression members of the truss. The vertical members (stirrups) make the vertical tensile stresses, while the steel reinforcement takes the horizontal tensile stresses on the bottom. Since its inception, many authors have modified this model, but the basics remain the same.

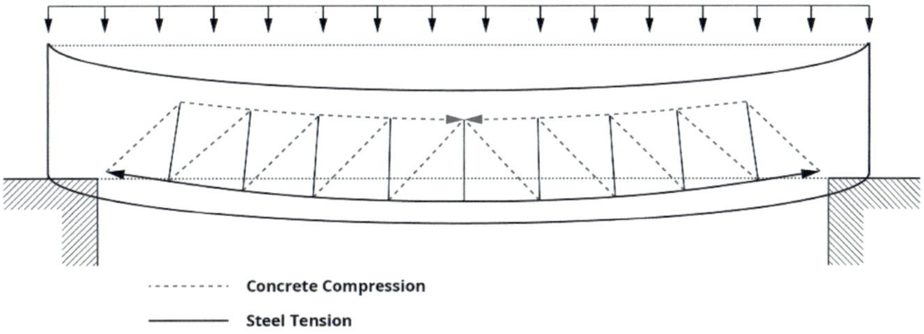

Figure 16. Ritter's original truss model (Kim, 2021).

The truss analogy method, also referred to as the strut-and-tie method, became popular for the analysis of reinforced concrete diaphragms (Schlaich et al., 1987; Bull, 2004; Moehle et al., 2010, Bull & Henry, 2014; Scarry, 2014, Nakaki 2000). Kamiya (1990, 1998) applied the truss method to timber diaphragms with openings. Kessel and Schönhoff (2001) derived the equivalent diagonal stiffness for light timber diaphragms. Moroder et al. (2014) further refined and modified the concept for use with light and mass timber diaphragms. Because reinforced concrete strut-and-tie models rely on the tensile strength of reinforcement bars and the compression capacity of the concrete, the arrangement of the strut and ties is not unique. For timber diaphragms, the size of the panels is known, and around the perimeter of each panel are fasteners with a known relative stiffness. One can therefore place the equivalent diagonals across each panel element, automatically defining the truss grid. Since the panel connections are the main source of diaphragm flexibility, they need to be accounted for in the calculation of the diagonal stiffness. Even though some assumptions required by the shear field analogy are not satisfied for mass timber diaphragms, the panels mainly work in shear, with longitudinal stresses only induced by local irregularities or discrete forces.

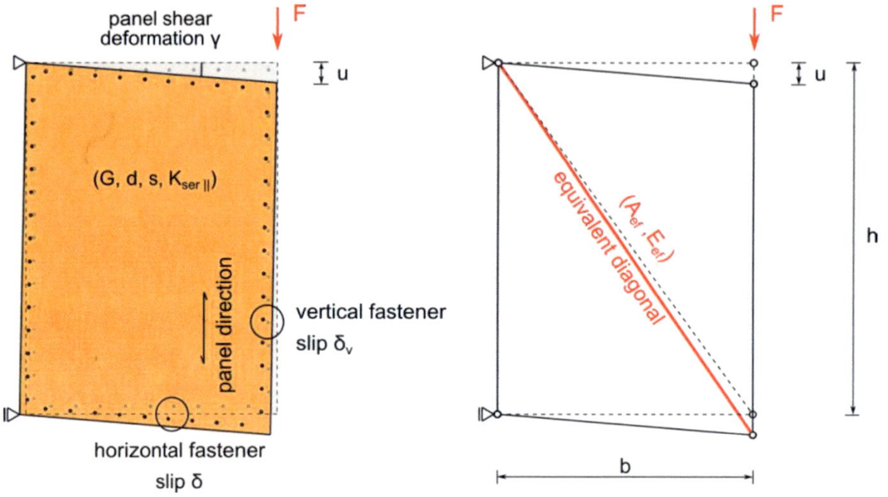

Figure 17. Shear panel with panel and fastener stiffnesses and equivalent truss diagonal (Moroder et al., 2014).

According to solid mechanics theory, as shown on the left side of Figure 17, the shear force F deforms a rectangular element into a parallelogram with the shear angle γ. The flexibility of the panel fasteners further increases the panel deflection. To account for this deflection, one can use a four-bar linkage system with an equivalent diagonal, as shown on the right side of Figure 17. The stiffness of the equivalent diagonal should give the same panel deformation under a given load F as the shear panel. Depending on the types of connections between the individual panels and between the panels and the beams, the equivalent shear-through-thickness rigidity of the panel $(Gd)_{eff}$ becomes

$$(Gd)_{eff} = \frac{1}{\left[\frac{1}{Gd} + \frac{2\,s}{K_{ser}}\left(\frac{1}{b} + \frac{1}{h}\right)\right]} \tag{3}$$

where

$(Gd)_{eff}$	is the equivalent shear-through-thickness rigidity of the panel;
G	is the shear modulus of the sheathing;
d	is the sheathing panel thickness;
s	is the fastener spacing;
K_{ser}	is the slip modulus of the fastener parallel to the panel edge;
b	is the panel width; and
h	is the panel height.

If the unit shear force (or shear flow) in the panel is set equal to the value of the axial stress σ then the equivalent cross-sectional area of the diagonal is numerically equal to the diagonal length, l:

$$A_{eff} = l = \sqrt{h^2 + b^2} \tag{4}$$

where A_{eff} is the equivalent cross-sectional area of the diagonal.

The equivalent modulus of elasticity of the diagonal E_{eff} is as follows:

$$E_{eff} = \frac{(Gd)_{eff}\, l^2}{h\, b} \tag{5}$$

Although setting the equivalent diagonal cross-sectional area A_{eff} as equal to the diagonal length l has no physical or mathematical meaning, it somewhat simplifies the method. The unit shear force v in the panel (the shear force per length) can be obtained as the normal stress in the diagonal:

$$v = \sigma = \frac{F_d}{A_{eff}} = \frac{F_d}{l} \tag{6}$$

where

v	is the unit shear force in the panel;
σ	is the axial stress in the diagonal;
F_d	is the force in diagonal; and
A_{eff}	is the diagonal area, which is equal to the diagonal length l.

To obtain the tension and compression forces in the chords and collector beams, one must add the integration of the unit shear forces along the element length to the axial forces from the truss elements. This is because the diagonal introduces the equivalent panel force in the nodes, whereas in reality it is introduced gradually through the fasteners along the panel edge.

Moroder et al. (2015) expanded this method for determining the effective shear-through-thickness rigidity to mass timber diaphragms. Depending on the type of connection between the individual panels and between the panels and the beams, the equivalent shear-though-thickness rigidity $(Gd)_{eff}$ for mass timber diaphragms is as follows:

$$(Gd)_{eff} = \frac{1}{\frac{1}{Gd} + \frac{s}{K_{ser}}\left(\frac{c_1}{b} + \frac{c_2}{h}\right)} \tag{7}$$

where

$(Gd)_{eff}$	is the equivalent shear-through-thickness rigidity of the panel;
G	is the shear modulus of the mass timber panel;
d	is the mass timber panel thickness;
K_{ser}	is the slip modulus of the fasteners parallel to the panel edge;
s	is the fastener spacing;
b	is the panel width;
h	is the panel height;
c_1	is the number of lines (rows) of fasteners between adjacent panels along the sheathing panel height h; and
c_2	is the number of lines (rows) of fasteners between adjacent panels along the sheathing panel length b.

Table 2. Values of the c_i coefficients for most used connections between diaphragm panels (Moroder & Chen, 2022)

Type of joint between diaphragm panels	*C* values
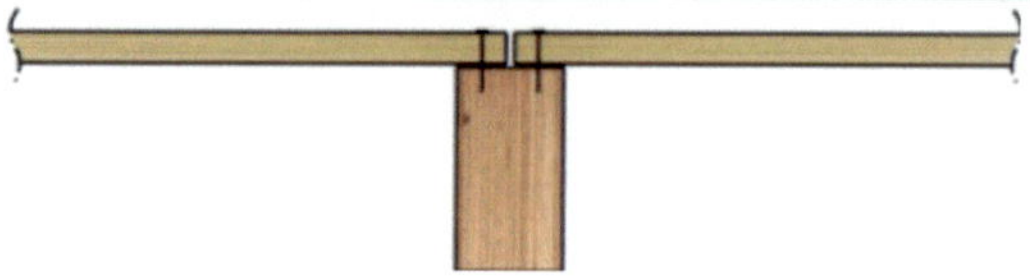 Light wood-frame diaphragm with nails	$c_1 = c_2 = 2$
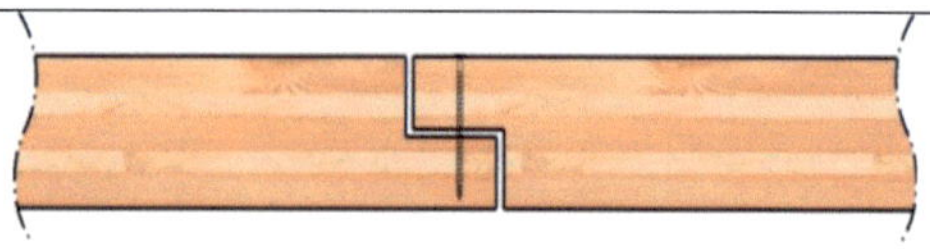 Mass timber diaphragm with lap joint and screws	$c_1 = 1$ $c_2 = 2$
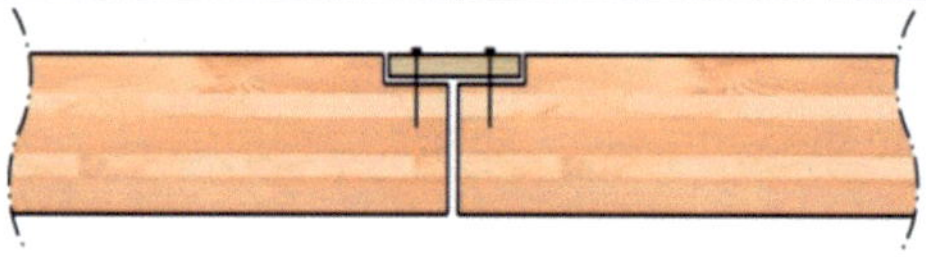 Mass timber diaphragm with single spline joint and screws	$c_1 = 2$ $c_2 = 2$

Values of the c_i coefficients for the most commonly used connections between panels are shown in Table 2. For blocked LWF diaphragms, the panels are connected via framing elements, requiring two lines of connections; therefore, $c_1 = c_2 = 2$. For mass timber panels, no framing elements are necessary along the longitudinal panel edge, so c_1 in this case is 1. If using a connection with a splice in such a case, $c_1 = 2$. The heads of the panels sit normally on a beam, requiring two lines of fasteners to transfer the forces; therefore, $c_2 = 2$. The effective area A_{eff} and modulus of elasticity E_{eff} of the equivalent diagonal can be derived with the same equations as for LWF diaphragms.

Mass timber panels normally span two gravity carrying beams and connect to them to guarantee diaphragm action. These beams are the truss elements running along the short length of the panels. There must be an equivalent truss element along the panel-to-panel edge, if there is no beam going in that direction. The stiffness of this element is the same as the panel stiffness in this direction, and one can assume the cross-sectional area is the sum of half the cross-section areas of each of the two panels. In this way, the truss analogy model can capture the longitudinal forces in the panel direction. Because mass timber panels possess relatively high axial stiffness compared to plywood/OSB panels, it is necessary to account for normal stresses along the two main directions. Additionally, fasteners will transfer forces not only parallel to the panel edges but also perpendicular to them. By dividing the panels into multiple diagonals, as shown in Figure 18, the transverse truss elements (along the panel width) can account for these effects by including the fastener stiffness perpendicular to the panel edges.

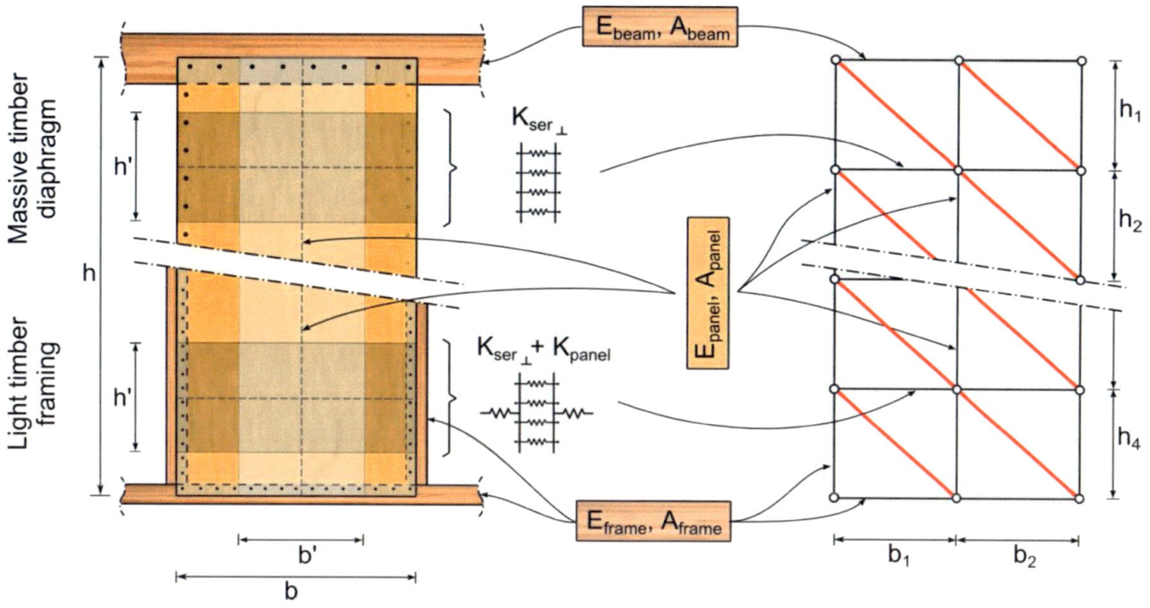

Figure 18. Idealization of wood-frame and mass timber diaphragm panel in a truss model with multiple diagonals (Moroder & Chen, 2022).

For panel subdivision into multiple diagonals, higher forces are attracted close to stiffer elements like beams or supports, due to their stiffness. In such case, the average of all diagonals belonging to one panel element should be considered. The individual diaphragm panel can be subdivided into a regular pattern, obtaining m x n equivalent diagonals, or by diagonals with varying lengths, as shown in Table 3 and Figure 19.

Table 3. Effective area and elasticity for regular and irregular truss diagonal models (Moroder & Chen, 2022)

Regular truss models (m x n)	Irregular truss models
$\boldsymbol{E_{eff,m\,x\,n} = mn\frac{(Gd)_{eff}l_{mxn}{}^2}{bh}}$	$E_{ef,ij} = \frac{(Gd)_{eff}l_{ij}{}^2}{b_j h_i}$
$\boldsymbol{A_{eff,mxn} = l_{m\,x\,n} = \sqrt{\left(\frac{b}{n}\right)^2 + \left(\frac{h}{m}\right)^2}}$	$A_{eff,ij} = l_{ij} = \sqrt{b_i{}^2 + h_j{}^2}$

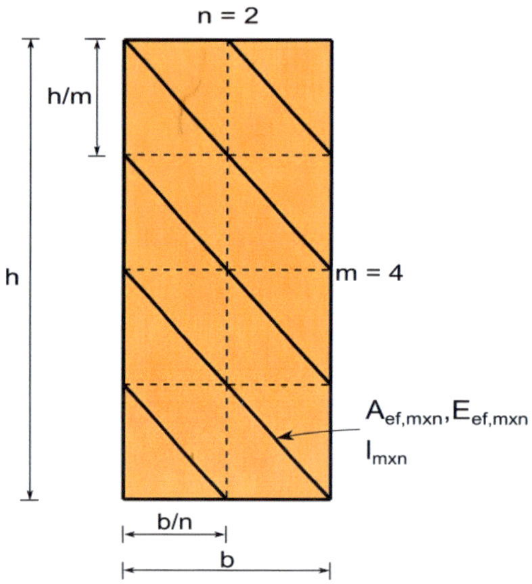

Figure 19. Multiple diagonals for regular m x n subdivision (left) or for irregular subdivision (right) (Moroder et al., 2014).

All beams (collector, chord, or strut) and framing elements, as well as other reinforcing elements, require modelling with their real axial stiffness. The remaining longitudinal truss elements (along the panel height h) should be modelled with an axial panel stiffness corresponding to the tributary width *b′* of the truss element, as shown in Figure 18. For the transverse truss element (along the panel width b) not corresponding to beams or framing elements, sum the stiffness of the tributary panel strip (in series) with the fastener stiffness perpendicular to the panel edge ($K_{ser\perp}$). Considering a common subdivision of two diagonals along the panel width, the equivalent stiffness (in force per length) of the transverse member for a LWF diaphragm can be calculated as shown in Equation (8):

$$K_{eff,tt} = \frac{1}{\dfrac{1}{\frac{E_{90}\,A'}{b'}} + \dfrac{1}{n'\,K_{ser,\perp}}} \tag{8}$$

where

$K_{eff,tt}$	is the equivalent stiffness of the transverse truss;
E_{90}	is the panel stiffness perpendicular to the panel direction;
A'	is the tributary cross section of the transverse truss element = $h'\,d$;
d	is the panel thickness;
b'	is the tributary width of the longitudinal truss element = $(b_i + b_{i+1})/2$;
h'	is the tributary width of the transverse truss element = $(h_i + h_{i+1})/2$;
n'	is the number of fasteners along h'; and
$K_{ser,\perp}$	is the slip modulus of the fasteners perpendicular to the panel edge.

For a mass timber diaphragm, the axial stiffness of the transverse truss element is much greater than that of the fasteners and can normally be ignored. For a common subdivision of two diagonals along the panel width, the equivalent stiffness of the transverse member in a mass timber panel can be calculated as follows:

$$K_{eff,tt,m} = \frac{1}{\frac{1}{\frac{E_{90}A'}{b'}} + \frac{1}{(3-c)\,n'\,K_{ser\perp}}} \cong (3-c)n'K_{ser,\perp} \tag{9}$$

where c is the number of fastener lines to transfer the unit shear force from one panel to the other.

3.2.3 Finite Element Modelling and Analysis

Numerical Finite Element (FE) Modelling can be a powerful tool for assessing diaphragm behaviour, especially in the assessment of the seismic response of a building. Using FE one can determine stresses in the panels and framing members, as well as forces in the connections and the deformation of all involved elements, under both monotonic and cyclic loading. The accuracy of the results is usually proportional to the model complexity, which again is related to the knowledge and time required to set up the model and post-process the results (Moroder & Chen, 2022). Most of the research on the computer modelling of timber diaphragms has been conducted for LWF diaphragms, starting with the pioneering work of Foschi and Bonac (1977) and Falk and Itani (1989). A series of standalone subroutines and analysis programs were developed over time to model the linear and nonlinear behaviour of wood-frame shear walls and diaphragms. For more information on most of these models, see Moroder and Chen (2022) and Popovski et al. (2022).

Commercially available software, like Dlubal's RFEM6 (Dlubal Software GmbH, 2016), SAP2000® (Computers and Structures Inc [CSI], 2016a), ETABS (CSI, 2016b), Abaqus (Dassault Systèmes, 2016), and ANSYS (Ansys Inc., 2023), as well as research software like OpenSEES (McKenna et al., 2000), can model mass timber structures and their diaphragm components. To successfully capture diaphragm behaviour, one must adequately model all relevant diaphragm components. Beam elements such as chords and collectors can be modelled with their real section sizes and material properties. The modelling of panel elements should be based on their thickness. Thinner panels, used more frequently in LWF construction, should be modelled as membrane elements by using the in-plane effective shear stiffness, as defined previously. Panel splice connections and connections between the panels and the beam elements need not be separately; they can instead be modelled together as one component along the length. The reduced shear stiffness of the diaphragm panels already considers the connection stiffness between the panel and beams.

In mass timber diaphragms, a single shell element with an equivalent shear stiffness that takes the panel splices into consideration, can be used for analysis of the load path in diaphragms and the load demand in the various diaphragm components. As timber panels have orthotropic properties, the modelling must include the shear strength and stiffness in the two orthogonal directions, taken either from the material standards or from the product manufacturer. For example, edgewise shear stiffness tests should be conducted in both the longitudinal (major) and transversal (minor) direction of a CLT panel, using the ASTM D198 (ASTM International, 2022) test method to determine the shear stiffness moduli $G_{e,0}$ and $G_{e,90}$, respectively, as per PRG 320 (American National Standards Institute & APA – The Wood Engineering Association [ANSI/APA], 2019). If given only the shear stiffness in the major direction, an approximation can help calculate the stiffness in the minor direction based on the geometry and panel layup given in Wallner-Novak et al. (2014). For simplicity, one can use the following approximation:

$$G_{e,90} = 0.75\, G_{e,0} \tag{10}$$

One should calculate the effective shear stiffness according to Equation ((10), using the values $G_{e,0}$ and $G_{e,90}$ for G in the equation. Some software packages, like Dlubal RFem (Dlubal Software GmbH, 2016), have specific modules that determine the orthotropic material properties of CLT based on the panel layup and laminate material properties.

The largest source of flexibility in timber diaphragms are the fasteners in the connections, and the in-plane stiffness needs to be reduced accordingly. This reduced or effective shear stiffness can be determined by introducing the slip modulus K_{ser} for the connections. This should be the elastic stiffness value of a specific connection or fastener, as, for example, provided in Eurocode 5 (European Committee for Standardization [CEN], 2004a). For a simplified analysis, it is not possible to use a nonlinear force-displacement curve as in some material standards, such as the New Zealand timber design standard (Standards New Zealand, 2022), or from test data. In these cases, one can consider a linearization with a secant stiffness taken at the yield point or at the design capacity of the fastener. This approach of determining a secant fastener slip modulus is a rough approximation of the fastener force-displacement behaviour, but it provides a workable value, considering the large uncertainties involved in determining the stiffness properties of fasteners in timber joints in general (Jockwer & Jorisen, 2018). Note that the effective shear stiffness of mass timber diaphragms is usually in the range of 20% to 40% of the timber-only shear stiffness. This range depends on the diaphragm geometry, as well as on the timber material and fastener selection.

When a more accurate diaphragm analysis is necessary, one must model the fastener stiffness with greater detail. This allows the designer to capture the behaviour of the various fasteners that connect the diaphragm components. The modelling of the panel-to-panel and panel-to-beam connections should include both their shear and axial stiffness. If the beam elements (chords or collectors) are spliced, then one must model the splice stiffness as well. The specification of the shell element should make use of the orthotropic material properties. One must also be careful to consider the correct orientation of the panel direction, as this will affect the orthotropic nature of the material.

Modelling the stiffness of the fasteners on the panel edge can provide a more realistic picture of the axial forces in the diaphragm. All simplified diaphragm analysis methods typically assume that it is the beam elements, rather than the sheathing panels, that resist any axial tension and compression forces. This behaviour is typically appropriate for a LWF diaphragm, but mass timber diaphragms attract significant axial forces. Engineering judgment is needed to decide the level of detail applied in modelling of the connections. Depending on the expected behaviour of the diaphragm (elastic or yielding), one should model the force-displacement behaviour of the connections with either linear or nonlinear springs or link elements. In the axial direction, a higher compression stiffness may be appropriate, to simulate the fact that the panels will be in contact once any tolerance gaps are closed. Depending on the software package used, continuous (line) hinges or discrete spring or link elements (Figure 20) can help simulate the connection behaviour.

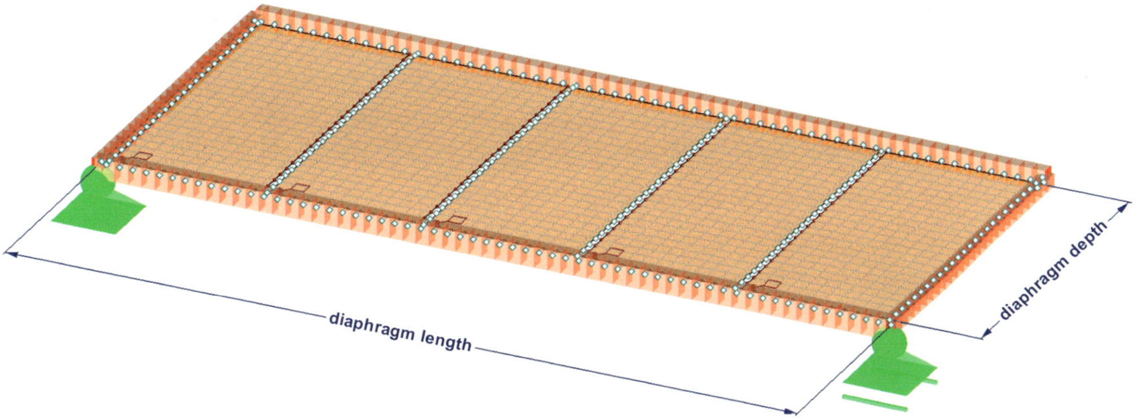

Figure 20. Diaphragm FE model with discrete fasteners (springs) used in model panel splices and connections of the panels to the chords and collectors (Moroder et al., 2022).

The latter allows for the direct modelling of additional fasteners or different types of fasteners in specific regions of the diaphragm, as well as the determination of any force concentrations in individual fasteners. Note that modelling the diaphragm in this way will take significant time, which many engineers may not be able to afford. As most software packages only allow for "orthogonal" stiffness properties, orientation of the resultant force and oriented spring properties can not be obtained. It is possible to read the axial forces in beam elements and fastener forces directly from the analysis results, but stresses in the shell element might require integration over the specific area under consideration (Moroder & Chen, 2022). Some software packages can represent force integrations over an ideal section through the shell element.

Breneman et al. (2016) discussed several approaches for modelling CLT diaphragms. As mass timber diaphragms use CLT, glulam, or SCL panels that are quite stiff in their own planes, most of the deflection of the diaphragms comes from the connections, and a significant portion of that deflection is located in the connection zones. Such behaviour also occurs in untopped precast concrete systems. The potential modelling approaches for CLT diaphragms fall into the following three categories of numerical models: (a) homogeneous, (b) discrete panel and lumped connections, and (c) discrete panel and distributed connections. Homogenous models are those whose panel-to-panel connection average is smoothed (averaged) over the entire model of the diaphragm (Figure 21a). This approach is similar to that used in LWF diaphragms, which verify combined diaphragm properties with experimental studies and implement them in national design standards. Given the many possible combinations of CLT diaphragm designs and proprietary connection types, significant research is needed to standardize homogeneous methods for analysis of CLT and mass timber diaphragms in general.

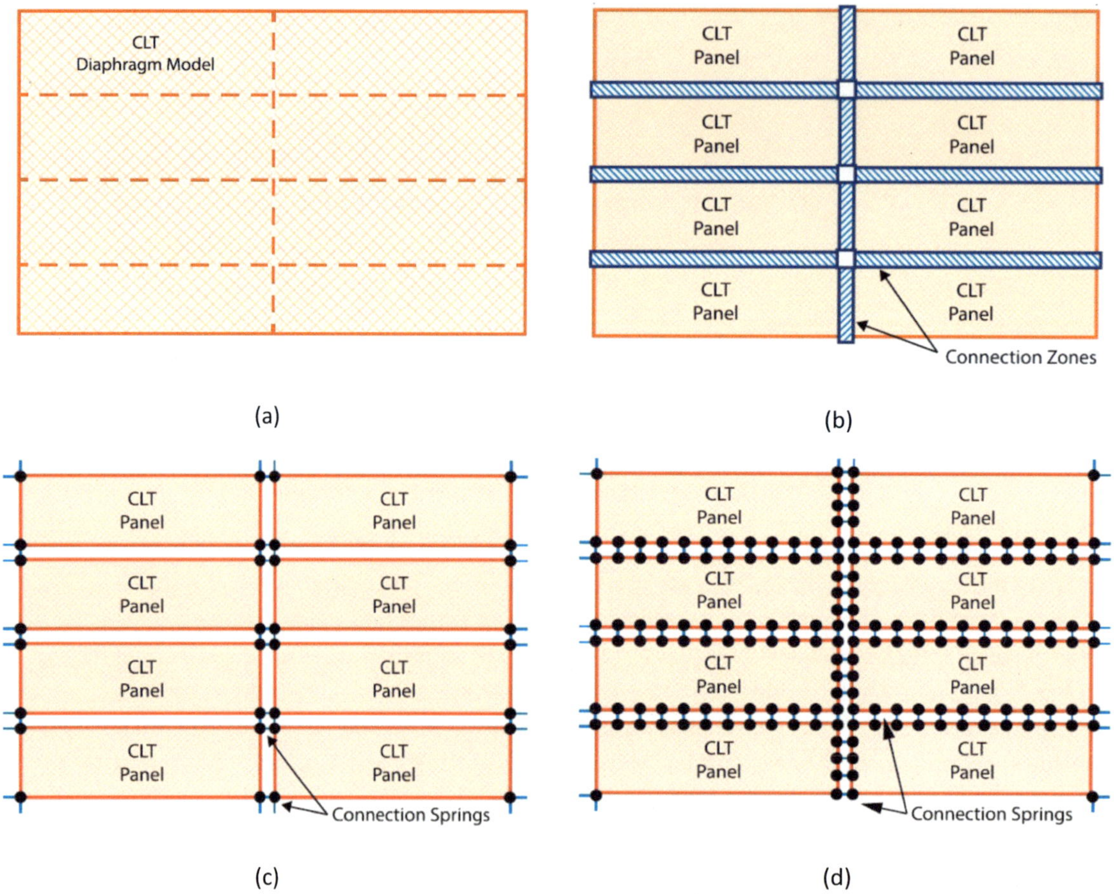

Figure 21. Different diaphragm models: (a) homogeneous model; (b) discrete model with 2-D connection zones; (c) discrete panel model with corner connections; and (d) discrete panel model with spaced connections (Breneman et al., 2016).

As previously mentioned in this section, discrete panel models explicitly model individual CLT panels, connections, or connection zones. The models may or may not include the connections to and between the diaphragm decking and the frame elements. Connection zone modelling can involve different levels of detail. One option, mostly used in research and/or product development, is to model each fastener with its linear or nonlinear force-deformation relationship. Breneman et al. (2016) have considered several intermediate, less detailed modelling approaches to be applied to real buildings. A distributed connection model includes representations of the connection zones as linear or 2-D modelling elements. For example, to represent a panel-to-panel connection, one can use shell or membrane elements with properties capturing the stiffness of the connection, both parallel and perpendicular to the panel edge, as shown in Figure 21b. Another discrete panel modelling approach is to aggregate the force-deformation behaviour of the connection zone to the ends of the panels using generalized two-point spring elements (Figure 21c). This approach has the advantage of including a relatively small number of elements but it may have a relatively large number of different element properties, particularly when there are many different panel sizes. Between the preceding two approaches is the option of discretizing the connection zones by placing two-point springs at regular spacing along each zone, as shown in Figure 21d. As such an approach requires the placement of a large number of elements, parametric generation of the analysis model using scripting, spreadsheets, or other model development tools can greatly aid the modelling process.

3.3 Diaphragm Stiffness and Deflection

3.3.1 Basics of Simplified Diaphragm Deflection

Over the last four decades, many studies have focused on the general topic of the stiffness and deflection of wood diaphragms. The first comprehensive publication on the deflection and design of timber diaphragms was the ATC Guideline for the Design of Horizontal Wood Diaphragms (Applied Technology Council [ATC], 1981), and those principles have been recently republished in Lawson et al. (2023). The ATC Guideline (ATC, 1981) contains some considerations regarding wood-frame diaphragm deflections and showed for the first time the derivation of the deflection based on first principles. It assumes that the diaphragm behaves like a deep I-beam (Figure 15 and Figure 22), in which the sheathing (generally plywood or OSB) acts as a web and resists the shear forces, while the diaphragm chords at the perimeter of the diaphragm act as flanges and resist the bending moments.

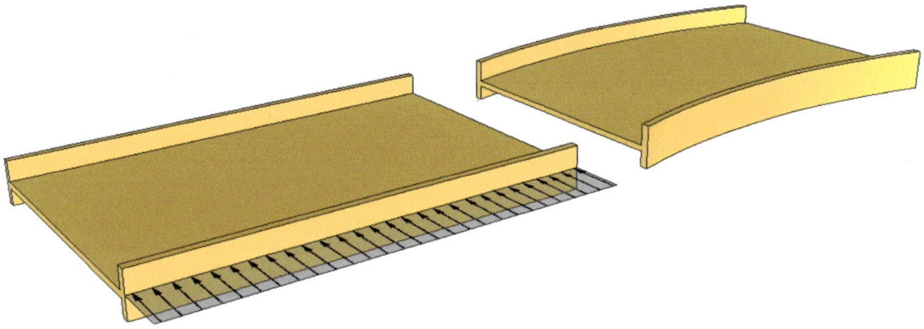

Figure 22. Idealized diaphragm represented as a deep I-beam.

In this case, the total deflection of a diaphragm can be attributed to four main contributions, as shown in Equation ((11): (a) Bending deformation of the chord beams; (b) Shear deformation of the sheathing panels; (c) Fastener slip between the panels and the framing members; and (d) Chord splice slip.

$$\Delta = \Delta_{bending} + \Delta_{shear} + \Delta_{slip} + \Delta_{splice} \tag{11}$$

Because of the limited available beam lengths, chords require splicing. Thus, the fourth term accounts for the splice slip on the diaphragm deflection.

This type of deflection equation, derived from first principles and supported by extensive testing on conventional wood-frame diaphragms, is still in use in a number of timber material design standards, including the Canadian Standard for Engineering Design in Wood, CSA O86 (Canadian Standards Association [CSA], 2019) and the U.S. Special Design Provisions for Wind and Seismic (American National Standards Institute & American

Wood Council [ANSI/AWC], 2021). According to section 11.7.2 of CSA O86-19, the deflection of a simply supported diaphragm can be determined using the following four-term Equation ((12):

$$\Delta_d = \frac{5vL^3}{96EAL_D} + \frac{vL}{4B_v} + 0.00061Le_n + \frac{\sum(\Delta_c x)}{2L_D} \tag{12}$$

where

Δ_d	is the lateral static deflection at mid-span, in mm;
v	is the maximum shear due to specified loads in the direction under consideration, in N/mm;
L	is the dimension of the diaphragm perpendicular to the direction of the load, in mm;
E	is the modulus of elasticity of the chord, in N/mm²;
A	is the cross-sectional area of the chord, in mm²;
L_D	is the dimension of the diaphragm parallel to the direction of the load, in mm;
B_v	is the shear-through-thickness rigidity of the sheeting, in N/mm;
e_n	is the sheathing-to-framing connection deformation, in mm; and
$\sum(\Delta_c x)$	is the sum of the individual chord-splice slip values, Δ_c, on both side of the diaphragm, each multiplied by its distance *x* to the nearest support.

Details of the derivation of the four-term Equation (12) appear in *Appendix A* of the *Guidelines for the Design of Horizontal Wood Diaphragms* (ATC, 1981). The following subsections examine each individual term to provide a clear understanding of the equation and how it can be adapted to mass timber diaphragms.

3.3.2 Bending Deformation Contribution

3.3.2.1 Light Wood-Frame Diaphragms

The derivation of the first term originates from the equation for a simply supported beam. The mid-span displacement for a simply supported beam under uniform load is defined as

$$\Delta_{B,mid\ span} = \frac{5wL^4}{384EI} \tag{13}$$

where w = the uniformly distributed load applied by the wall along the diaphragm length, N/mm.

Since this neglects the bending stiffness given by the sheathing, as shown in Figure 23, one can obtain the inertia of the chords, I, using the parallel axis theorem, as follows:

$$I = \sum(I_c + Ar^2) \tag{14}$$

If the individual inertia of each chord is neglected (I_c), the result is the following equation:

$$I = 2A\left(\frac{L_D}{2}\right)^2 = \frac{AL_D^2}{2} \tag{15}$$

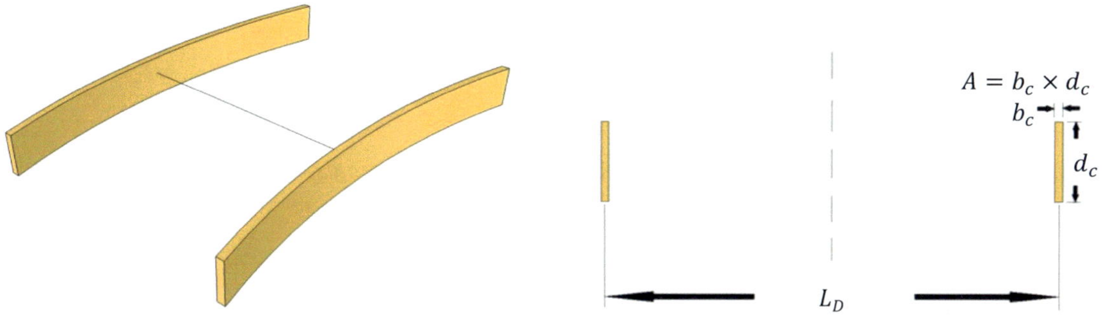

Figure 23. Simplified diaphragm cross section used to determine bending displacement.

The distributed load can then be rewritten in terms of the maximum unit shear at the support as follows:

$$w = \frac{2L_D v}{L} \tag{16}$$

Finally, substituting equations (15) and (16) into (13), one obtains the following equation:

$$\Delta_{B,mid\ span} = \frac{5vL^3}{96EAL_D} \tag{17}$$

3.3.2.2 Mass Timber Diaphragms

The bending term for the LWF diaphragm equation assumes that the sheathing elements, generally plywood or OSB, do not significantly contribute to the flexural rigidity of the diaphragm; instead, the flexural rigidity of the tension and compression chords is the most significant contribution to the bending capacity of the diaphragm. This simplification is based on the rationale that due to thinness of the panels, their deformation mechanisms are more closely related to fastener slip, shear deformation, and possibly buckling than to bending. For decking made of mass timber products, the connections between panels and between the panels and the framing are much stiffer, while the panels are much thicker. Consequently, the flexibility given by the connection is reduced and the bending stiffness the panels provide is increased, which overall increases the bending stiffness contribution of the panel.

Based on these observations, the bending stiffness of the decking may be included when evaluating the flexural stiffness of the mass timber diaphragm. However, unless there is no discontinuity, which will require very large panels or LLRSs very near each other, the connection flexibility may not allow a full engagement with the whole depth of the diaphragm in bending and will thus require the calculation of an effective bending stiffness. Furthermore, it could also be possible to integrate the connection flexibility into the calculation of the effective bending stiffness.

3.3.3 Shear Deformation Contribution

3.3.3.1 Light Wood-Frame Diaphragms

Beam analyses generally ignore shear displacements because they are normally negligible in comparison to bending deformations. The exception is for beams with small length-to-depth ratios (deeper beams), so when considering the diaphragm as a deep beam, the shear deflections shall be considered. Unlike bending deformations, shear deformations are controlled by the diaphragm's panels (web).

For a simply supported beam, the shear deformation under uniform load is

$$\Delta_{S,mid\ span} = \frac{wL^2\alpha}{8A_pG} \tag{18}$$

where

$\Delta_{S,mid\ span}$	is the lateral static deflection due to shear deformation at mid-span, in mm;
G	Is the shear modulus, in N/mm^2;
α	is the form factor accounting the for the shear stress distribution. $\alpha = 1$, for the diaphragm's panels;
A_p	is the cross-sectional area of the panel =$L_D \times t$, in mm^2; and
t	is the panel thickness, mm.

Substituting Equation (16) into (18) and assuming $\alpha = 1$ and $A_p = L_D \times t$ gives

$$\Delta_{S,mid\ span} = \frac{vL}{4tG} \tag{19}$$

Finally, defining the shear-through-thickness rigidity of the sheathing, B_v, as the product of the panel thickness, t, and the shear modulus, G, provides the second term of Equation (12) for the shear deformation of the diaphragm:

$$\Delta_{S,mid\ span} = \frac{vL}{4B_v} \tag{20}$$

3.3.3.2 Mass Timber Diaphragms

The shear deformation for LWF diaphragms depends on the shear-through-thickness rigidity of the panels, which is in turn proportional to the product of the panel's shear rigidity and its thickness. The shear stresses induced by the in-plane forces result in shear deformations, as shown in Figure 24.

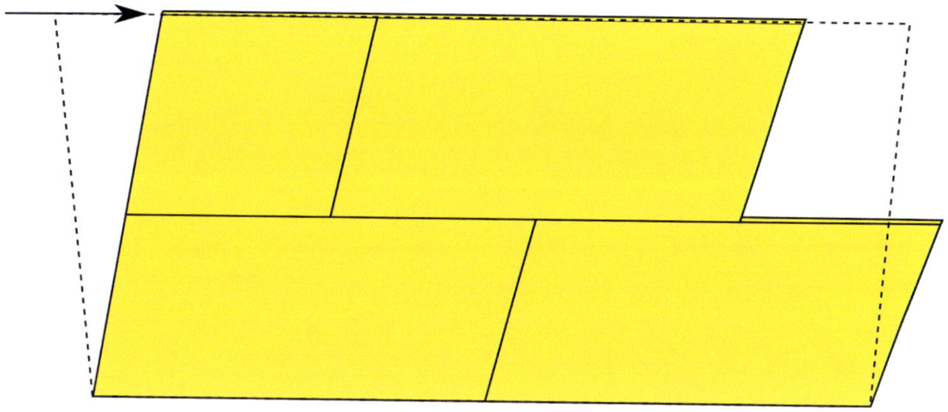

Figure 24. In-plane shear deformations of LWF diaphragms.

For mass timber panels and decking, the same principles still apply. The apparent shear stiffness could include the flexibility of the connection between panels. More information on this topic is presented in the fastener slip section.

3.3.4 Fastener Slip

3.3.4.1 Light Wood-Frame Diaphragms

Appendix A of the *Guidelines for the design of horizontal wood diaphragms* (ATC, 1981) discusses deflection attributable to nail slip. The theory is based on the assumption that fastener displacements under load are analogous to web shear. Under this assumption, forces along an individual panel's edges are due to the shear forces and must be equal for both parallel- and perpendicular-to-load directions. One can derive the fastener slip contribution in terms of nail deformations (e_n) in the directions of these shear displacements. It is possible to evaluate the effects of these deformations separately at the mid-length of the diaphragm, as represented by $\Delta_{\parallel}$ and $\Delta_{\perp}$ in Figure 25 and Figure 26, respectively. Note that the schematic resented in these figures and in the following equations assume a constant fastener slip along the span of the diaphragm. In reality, a larger slip is expected at the support and smaller slip at the mid-span. Deflection attributed to nail-slip is then as follows:

$$\Delta_{fastener\ slip,mid\ span} = \Delta_{\parallel,mid\ span} + \Delta_{\perp,mid\ span} \tag{21}$$

Following the schematic shown in Figure 25, the mid-span deflection of the diaphragm due to the nail-slip in the direction parallel to the load can be defined as follows:

$$\Delta_{\parallel,mid\ span} = \frac{e_n L}{2L_p} \tag{22}$$

where L_p is the length of an individual panel.

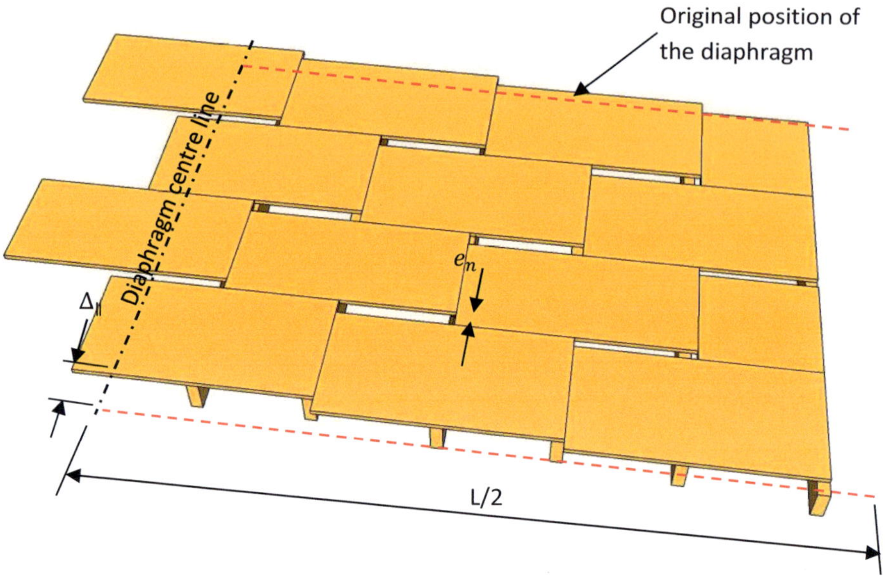

Figure 25. Fastener slip deflection contribution parallel to the applied load.

Following Figure 26, which shows a rotation of $\frac{e_n}{W_p}$ at the support location, the diaphragm mid-span deflection due to the nail slip in the direction perpendicular to the load is as follows:

$$\Delta_{\perp,mid\ span} = \frac{e_n}{W_p}\frac{L}{2} \tag{23}$$

where W_p is the width of an individual panel. Substituting equations (22) and (23) into Equation (21) gives

$$\Delta_{fastener\ slip,mid\ span} = \frac{e_n L}{2}\left(\frac{1}{L_p} + \frac{1}{W_p}\right) \tag{24}$$

Since Equation (12) was derived for standard 4 by 8 ft. (1220 x 2440 mm) plywood or OSB sheathing panels, one can substitute $W_P = 1220$ mm and $L_P = 2440$ mm into Equation (24) to obtain

$$\Delta_{fastener\ slip,mid\ span} = 0.00614 e_n L \tag{25}$$

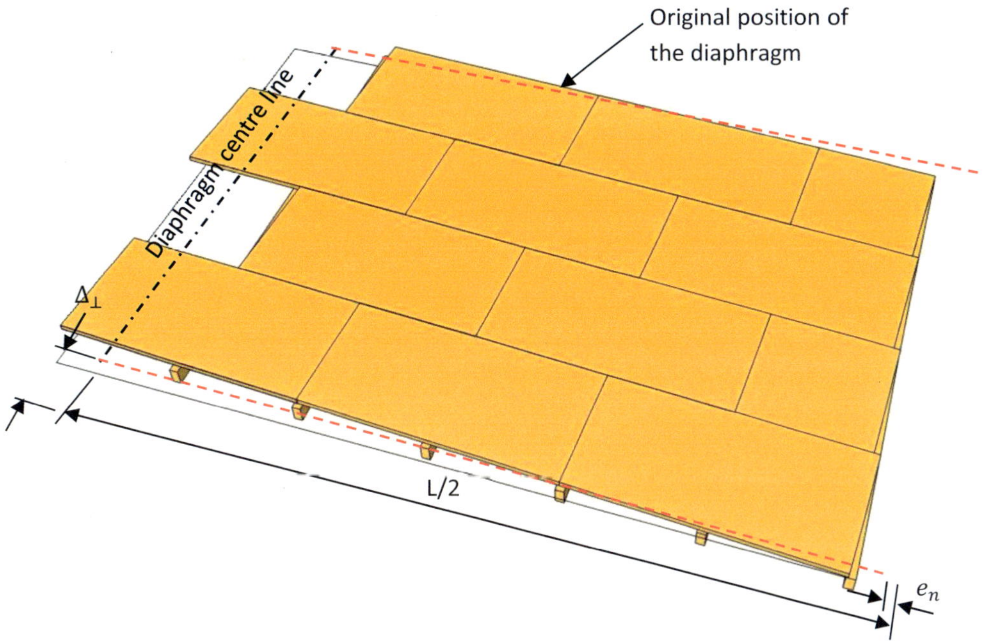

Figure 26. Deflection contribution from fastener slip perpendicular to the applied load.

In its annexe (clause A.11.7), CSA O86-19 defines the following equation to estimate the fastener slip in a diaphragm or shear wall:

$$e_n = \left(\frac{0.0013vs}{d_F^2}\right)^2 \tag{26}$$

where

- d_F is the nail diameter;
- s is the spacing of the nail along the panel edge; and
- v is the maximum specified shear force per unit of length along the diaphragm boundary.

3.3.4.2 Mass Timber Diaphragms

The fastener slip deflection for LWF diaphragms is based on the geometry of standard sheathing panels (i.e., 4 x 8 ft.) and the stiffness of the panel-to-frame fasteners. In mass timber diaphragms, not only is the aspect ratio significantly different, but both the panel-to-frame and panel-to-panel assemblies use mechanical fasteners. Consequently, the simple equation developed for LWF diaphragm is not suitable for a mass timber diaphragm. Moreover, one can consider the contribution of the fastener slip by estimating an apparent bending stiffness and an apparent shear stiffness.

Before determining the fastener slip contribution to the deflection, one must define the slip mechanism. As with a LWF diaphragm, the slip can be either parallel (Figure 6b) or perpendicular to the load (Figure 6a). In both cases, to estimate the fastener slip contribution, one can reduce the apparent shear stiffness, since this kind of slip is induced by shear load. If there are several discontinuities along the diaphragm span and width, the following equation can estimate an effective shear modulus:

$$G_{eff} = \left(\frac{1}{G} + \frac{s \cdot t}{k}\left(\frac{1}{W_p} + \frac{1}{L_p}\right)\right)^{-1} \tag{27}$$

where

G	is the shear modulus of the panel in N/mm²;
s	is the spacing of the connection at the panel perimeter in mm;
t	is the panel thickness in mm;
k	is the connection stiffness in N/mm;
W_p	is the panel width; and
L_p	is the panel length.

However, since the panel can take bending moment, it becomes necessary to also define a different type of fastener slip deformation. As shown in Figure 27, a rotation between panels is possible. This type of deformation is similar to a chord slip that induces an increment of rotation.

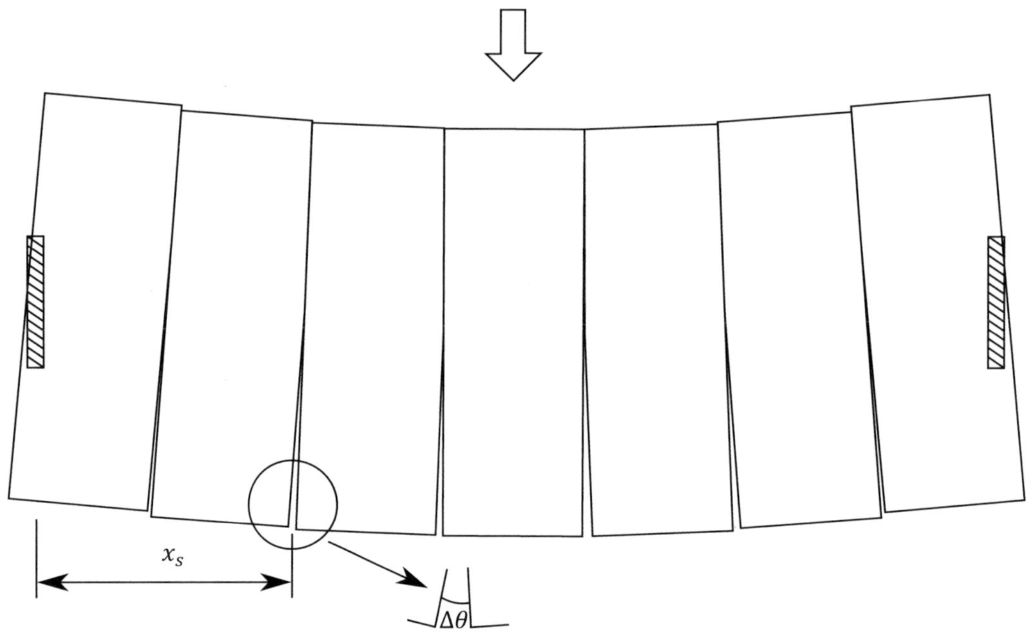

Figure 27. Fastener slip due to bending deformation.

The increment of rotation caused by the discontinuity shown in Figure 27 in turn causes an increment in the deflection. One can determine that increment in deflection at mid-span using the following equation for a single span system:

$$\Delta_{mid\ span} = \sum \frac{x_s}{2} \Delta\theta \tag{28}$$

where x_s is the distance between the analyzed discontinuity and the nearest support and $\Delta\theta$ is the increment of rotation at the analyzed discontinuity. That increment of rotation is equal to the following equation:

$$\Delta\theta = \frac{M}{k_\theta} \tag{29}$$

where M is the moment applied at the analyzed discontinuity and k_θ is the rotational stiffness from the connection. If the connection is spaced continuously through the depth of the panel, k_θ is equal to the following equation:

$$k_\theta = \frac{kL_D^3}{12s} \tag{30}$$

where

k	is the connection stiffness of a single connector;
s	is the connection spacing along the panel depth; and
L_D	is the panel depth.

Note that Equation (30) assumes a uniformly distributed stiffness along the panel depth. However, even with large spacing, the equation remains sufficiently accurate. For example, even with as few as six connectors along the panel depth, Equation (30) will overestimate the stiffness by only about 3%.

If there are several discontinuities along the diaphragm span, instead of evaluating the increment of deflection caused by these discontinuities, one can reduce the bending stiffness to the value in the following equation:

$$(EI)_{eff} = \left(\frac{1}{EI} + \frac{12s}{kL_D^3 L_p}\right)^{-1} \tag{31}$$

where EI is the bending stiffness of the panel and the chord, assuming a rigid connection between the components, and L_p is the panel length in the span direction.

3.3.5 Chord Spice Split

3.3.5.1 Light Wood-Frame Diaphragms

Appendix A of the *Guidelines for the Design of Horizontal Wood Diaphragms* (ATC, 1981) covers deflection attributed to chord splice deformations. It follows the assumption that the rotation of the diaphragm at the supports is proportional to individual chord slip (Δ_c) and distance from the splice location to the nearest support (x_s).

The deflection created by the chord splice is due to the localized increase in rotation, as shown in Figure 28. The following equation gives the total increase in rotation at the chord splice:

$$\theta = \theta_1 + \theta_2 = \frac{\Delta_c}{L_D} \tag{32}$$

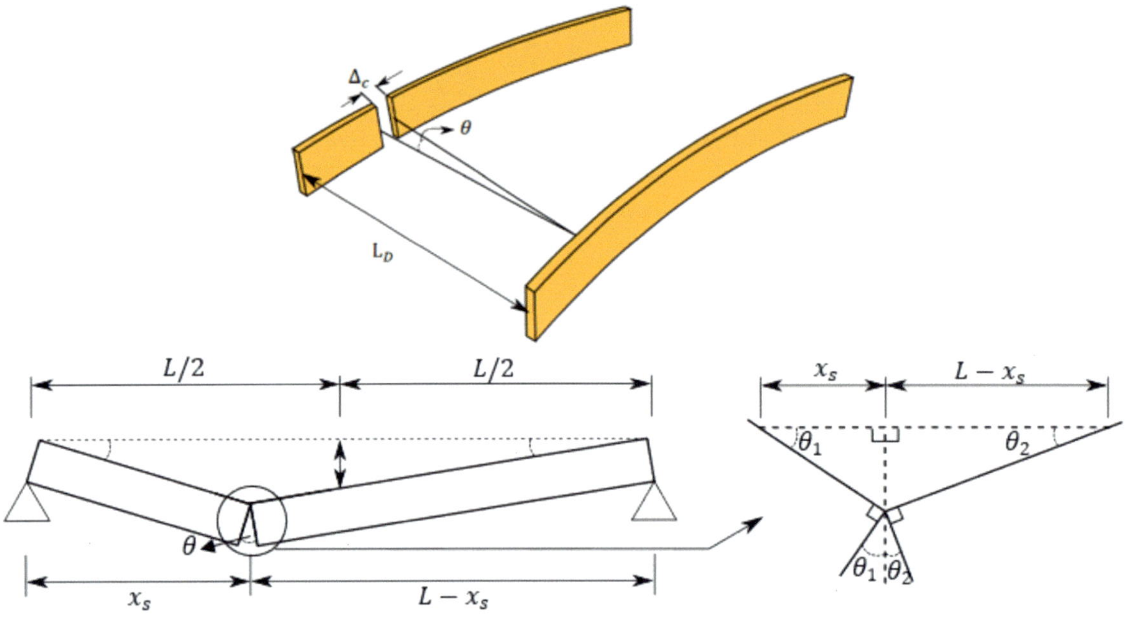

Figure 28. Deflection of a diaphragm due to chord splice slip.

The localized rotation from the chord splice slip induces rotation at the support, as shown in Figure 28. From trigonometry and the approximation of small angles, the rotation at the farthest support for the chord splice is

$$\theta_1 x_s = \theta_2 (L - x_s) \tag{33}$$

$$\theta_2 = \frac{x_s}{L}\theta = \frac{x_s}{L}\frac{\Delta_c}{L_D} \tag{34}$$

The mid-span deflection due to the chord slice slip is

$$\Delta_{splice,mid\ span} = \theta_2 \frac{L}{2} = \frac{\Delta_c x_s}{2L_D} \tag{35}$$

When considering multiple spices, Equation (35) becomes

$$\Delta_{all\ splice,mid\ span} = \frac{\sum \Delta_c x_s}{2L_D} \tag{36}$$

3.3.5.2 Mass Timber Diaphragms

Mass timber products allow for longer members than the lumber used in the chord construction of LWF diaphragms. However, the span between the vertical LLRS in mass timber construction is generally larger than in LWF construction. This implies that mass timber diaphragms could require a similar quantity of chord splices. For LWF diaphragms, the assumption is that the chord entirely takes the bending load as per Equation ((36).

Using the same equation for mass timber diaphragms should be conservative because the decking can normally carry a part of the bending load, which somewhat restrains chord splice slips.

4 STIFFNESS AND SLIP OF MASS TIMBER CONNECTORS: CODES AND STANDARDS PERSPECTIVE

The stiffness of mass timber diaphragms depends on the in-plane stiffness of the mass timber plate elements, the stiffness of the connections between them, and the stiffness of other elements, such as diaphragm chords and collectors. As there are connections not only present between the mass timber plate elements but also in other parts of the diaphragm, they can be among the most important parameters in quantifying the diaphragm stiffness and its performance. This section discusses how recognized codes and standards deal with initial connection stiffness.

4.1 Connection Stiffness in Testing Standards

The load-deformation behaviour of mass timber connections with dowel-type fasteners is nonlinear and highly complex (Foschi & Bonac, 1977; Dolan & Foschi, 1991; Heine & Dolan, 2001; Lemaître et al., 2018). An example of a ductile load-deformation curve for a timber connection tested under monotonic loading using the EN 26891 standard (CEN, 1991) appears in Figure 29a. The initial part of the curve is characterized by low stiffness at low load levels, mainly due to tolerances in the assembly of the connection for fasteners like bolts or dowels (initial slip in Figure 29a).

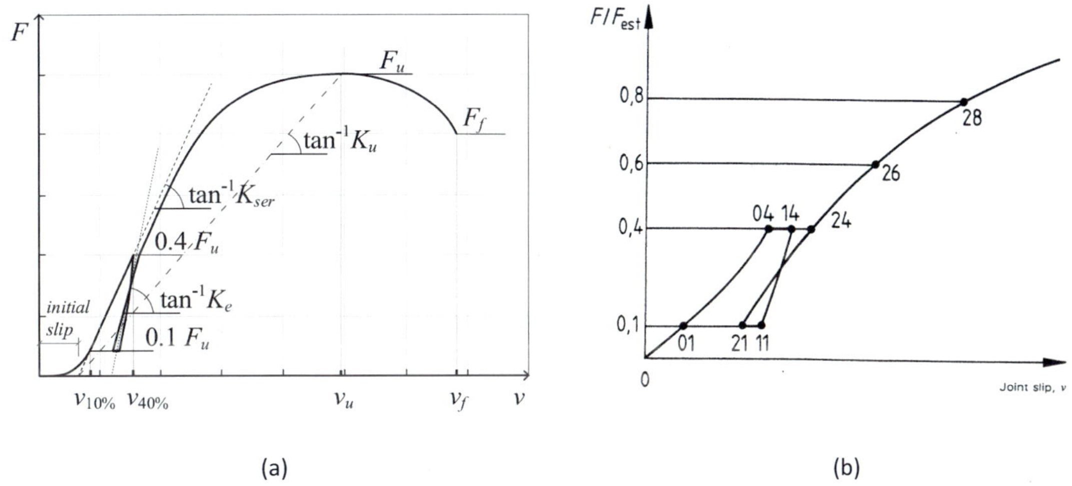

Figure 29. (a) Load-deflection curve of a typical connection tested using the EN 26891 standard (Jockwer et al., 2021) and (b) the idealized load-deformation procedure according to EN 26891.

This softness in behaviour can be more pronounced in real applications due to the higher precision of laboratory manufacturing as compared to connections produced in practice. The soft part of the curve is not as pronounced in fasteners such as nails and STSs. With the latter, there is sometimes a high initial stiffness. Depending on the thread design and length, strong contact can develop between the material surfaces, generating friction at the beginning of the loading. When the components of the connection are in full contact, the load-deformation behaviour becomes approximately linear. This linear range is typically between 10% and

40% of the maximum load, so most testing standards involve either a part or the entire load section when calculating the elastic stiffness (slip modulus) of the connection. The same range applies for calculation and modelling purposes in the serviceability limit states of connection performance.

This slip modulus, designated as K_{ser} in Figure 29 according to Eurocode 5 (EN, 2004a), can be calculated according to EN 26891 (CEN, 1991) as

$$K_{ser} = \frac{0.4 \cdot F_u}{\frac{4}{3}(v_{0.4} - v_{0.1})} \tag{37}$$

where F_u is the maximum load, also designated as F_{max} in the literature; $v_{0.4}$ and $v_{0.1}$ are the deformations corresponding to 40% and 10% of F_u, respectively.

When performing connection tests according to EN 26891 with an unloading and reloading cycle between 40% and 10% of the estimated maximum load, one can determine the elastic slip modulus K_e according to Equation (38) using the deformations v_{14} and v_{11} at 40% and 10% of F_u, respectively, for the unloading cycle, along with deformations v_{21} and v_{24} at 10% and 40% of F_u, respectively, for the reloading cycle.

$$K_e = \frac{0.4 \cdot F_u}{\frac{2}{3}(v_{14} + v_{24} - v_{21} - v_{11})} \tag{38}$$

Note that there is a difference between the tangential slip modulus (stiffness) of the load-deformation curve at a single point and the secant slip modulus, which derives from the differences in load and deformation at two points along the curve. The standard definition of the slip modulus for the ultimate limit state K_u is usually defined as the secant between zero and the ultimate load, as shown in Figure 29, or that between 0% and 60% of the ultimate load (Cuerrier-Auclair, 2020).

For the cyclic testing of connections according to ISO 16670 (ISO, 2003) or Method B of ASTM E2126 (ASTM International, 2019), the elastic stiffness of the connection should be

$$K_e = \frac{0.3 \cdot F_{max}}{(v_{40\%} - v_{10\%})} \tag{39}$$

where F_{max} is the maximum load, while $v_{40\%}$ and $v_{10\%}$ are deformations corresponding to 40% and 10% of F_{max}, respectively.

The Swiss standard for the design of timber structures, SIA 265 (Swiss Society of Engineers and Architects [SIA], 2012), defines the slip modulus K_{ser} in the serviceability limit state as the ratio between the yielding load of the connection F_y and the yield deformation v_y, as shown in Equation (40). This approach is also more directly related to the calculation of elastic stiffness in steel and concrete connections and components.

$$K_e = \frac{F_y}{v_y} \tag{40}$$

The drawback of using this approach for timber connections is that there is no agreed-upon method to define the yield displacement of the connection.

4.2 Determination of the Yield Deformation and Yield Load

The yield point of a connection is typically defined as the load at which the connection begins to experience irreversible plastic deformations. This is usually detectable from the load-displacement curve that is obtained from testing. However, the yield point in timber-based connections is not as clearly defined as in connections from other structural materials. Consequently, there are different methods for defining this yield point in timber connections. This section briefly describes the widely accepted, yet different, methods for determining the yielding deformation based on load-displacement curves obtained from monotonic or cyclic tests on connections.

The European standard EN 12512 (CEN, 2001) that regulates how to conduct cyclic tests on timber joints states that the preferred method for determining the yield point should be the load-displacement curve obtained from monotonic tests conducted according to EN 26891 (CEN, 1991), following one of two approaches: (a) If the load-displacement curve covers two well-defined linear parts, the yield points (load and displacement) are defined by the intersection between these two lines (Figure 30 left); (b) for all other cases, one should determine the yield point as shown in Figure 30 right. In the second option, the slope of the secant line connecting the two points on the curve, from 0.1 F_{max} to 0.4 F_{max} (angle α in Figure 30), serves as the initial stiffness K_i. The yield point is determined by the intersection of the initial stiffness line and the tangent line that has a stiffness 1/6 that of the initial value, as shown in Figure 30 right; $\tan(\beta) = 1/6 \tan(\alpha)$. EN 12512 also defines the ultimate load F_u as corresponding to the point of failure, which is either 80% of F_{max} or the load occurring at a displacement of 30 mm, whichever occurs first. This affects the calculation of ductility and is not related to the initial stiffness.

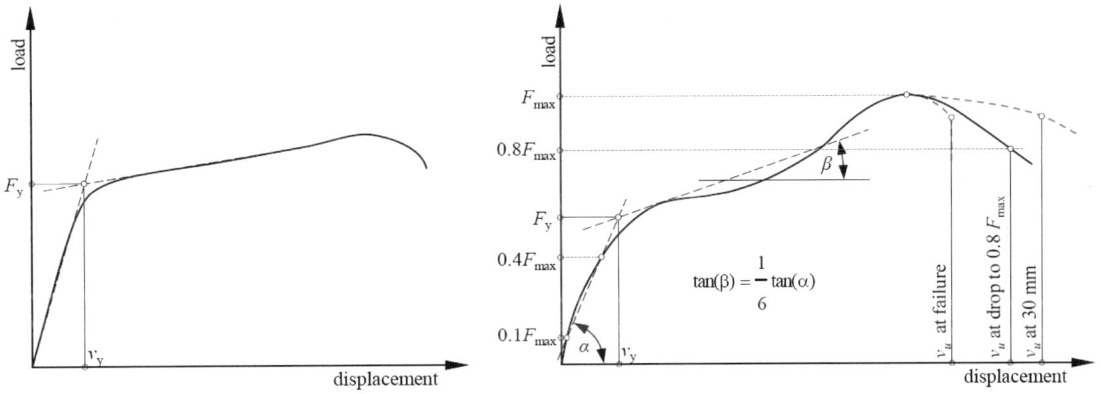

Figure 30. Definition of the yield and the ultimate point according to EN 12512 for two well-defined linear parts (left); and when the test curve does not cover well-defined linear parts (right).

ISO 16670 (ISO, 2003) is the international equivalent to the EN 12512 standard, and it suggests to calculate the stiffness in the same way. The Swiss SIA 265 standard (SIA, 2012) includes a definition for the yield point similar to the second method described in EN 12512. The only difference is that it defines the initial stiffness as the line between the origin and 0.4 F_{max} (Figure 31).

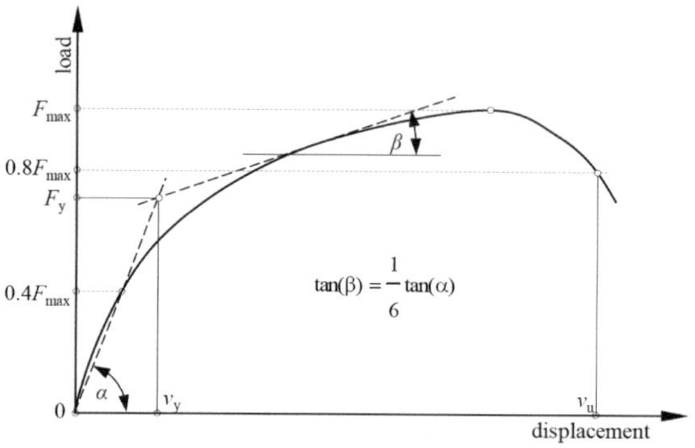

Figure 31. Definition of the yield point according to SIA 265.

The American standards ASTM D5652 (ASTM International, 2021) and ASTM D5764 (ASTM International, 2023) define the proportional limit and yield displacement of a single-bolt connection in wood and wood-based products and evaluate the dowel-bearing strength in these products, respectively. They use a 5% offset line (Figure 32) to determine the yield point. In other words, the standard requires to horizontally offset the line representing the initial stiffness for 5% of the fastener's diameter.

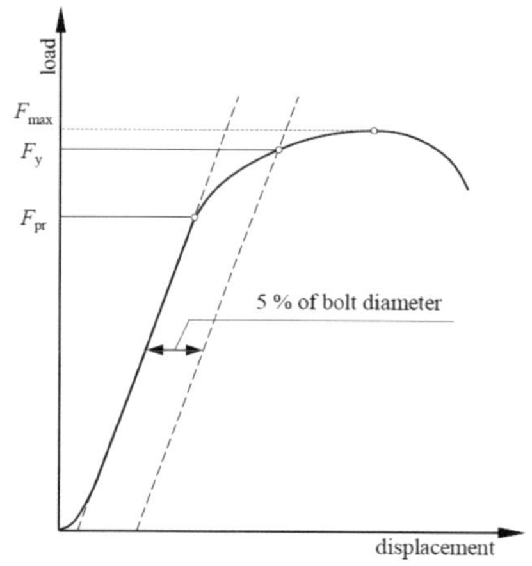

Figure 32. Definition of the yield point according to ASTM D5652 and D5764.

The equivalent energy elastic-plastic (EEEP) model is one of the methods defined in ASTM E2126 (ASTM International, 2019), and one of the most widely used in North America. As the name suggests, this model establishes an elastic-plastic approximation around the load-displacement curve through the equal energy dissipation rule. In other words, it approximates the load-displacement curve with a bilinear model so that the area below the elastic-plastic model is the same as the area below the curve. The initial stiffness of the bilinear model is calculated between the origin and 0.4 F_{max}, or P_{peak}, as noted in the standard, while displacement at failure Δ_u occurs at the descending part of the load-displacement curve at 80% of the maximum load P_{peak} (Figure 33).

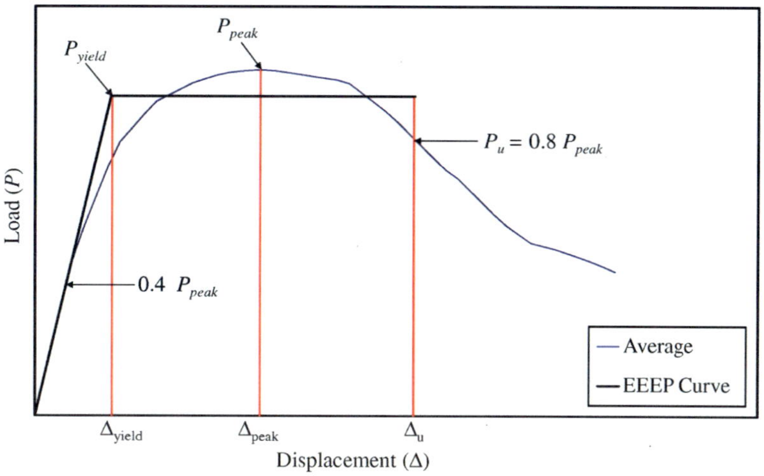

Figure 33. Definition of the yield point according to ASTM D5652 and D5764.

One can calculate the yield load P_{yield} according to Equation (41):

$$P_{yield} = \left(\Delta_u - \sqrt{\Delta_u^2 - \frac{2A}{K_e}} \right) K_e \tag{41}$$

where A is the area under the load-displacement curve until Δ_u and K_e is the initial elastic stiffness. The intersection of the line representing the initial stiffness and a horizontal line fixed at P_{yield} defines the yield point, where one can determine the corresponding yield displacement.

The literature proposes several other methods for defining the yield displacement and initial stiffness, but for brevity, these are not included in this section. Those include Ehlbeck (1979), Yasumura and Kawai (1998), Karacabeyli and Ceccotti (1996), and Foliente (1996), also referred as the CSIRO Method.

4.3 Stiffness of Connections in Design Standards

The stiffness of connections in timber diaphragms is one of the most important parameters influencing their performance under in-plane loads. As connection behaviour is highly nonlinear (Figure 29), there have been different ways of describing their initial stiffness, as discussed in the previous section. This section explains how various national standards have dealt with the stiffness of connections.

4.3.1 *Eurocode 5*

Eurocode 5 (EC5) refers to European Standard EN 26891 to determine the slip modulus K_{ser} in the serviceability limit state and gives formulas for different types of fasteners. For fasteners such as dowels, bolts, screws, or predrilled nails, EC5 suggests that $K_{ser} = \rho_m^{1.5} \cdot d/23$, where ρ_m is the mean material density and d is the fastener diameter. For nails without predrilling, it recommends $K_{ser} = \rho_m^{1.5} \cdot d^{0.8}/30$. For staples $K_{ser} = \rho_m^{1.5} \cdot d^{0.8}/80$ should be used. For wood-to-steel or wood-to-concrete connections, the standard states that the stiffness should be calculated based on the density of the wood member and may be multiplied by 2.0. The standard, however, makes no distinction between fasteners loaded parallel or perpendicular to the grain.

When the mean densities ρ_{m1} and ρ_{m2} of the two jointed wood-based members are different, ρ_m in the above formulas can be assumed as the square root of the product of both values: $\rho_m = \sqrt{\rho_{m1}\,\rho_{m2}}$.

The background for these equations appears in Ehlbeck and Larsen (1993) and are based on the ratio of the load and deformation at 40% of the characteristic load-carrying capacity according to the European yield model. In this ratio, to derive the load, one inserts the equations for embedment strength and yield moment into the European yield model equations while neglecting all partial safety factors for a nail in single shear in the failure mode with two plastic hinges per shear plane (Jockwer et al., 2021). The deformation at 40% of the characteristic load-carrying capacity is based on studies by Ehlbeck and Werner (1988a and 1988b) and is as follows:

$$v_{40\%} \approx \frac{40 \cdot d^{0.8}}{\rho_k} \tag{42}$$

where ρ_k is the characteristic density. The resulting equation for the slip modulus of a connection with predrilled nails of diameters 2 mm ≤ d ≤ 8 mm depends on the diameter and characteristic value of the density of the timber members:

$$K_{ser} = \frac{0.4 \cdot F_{v,Rk}}{v_{40\%}} = \frac{0.55}{100}\sqrt{100 - d\,\rho_k^{1.5}} \approx \frac{\rho_k^{1.5}\,d}{20} \tag{43}$$

The relation above was incorporated in the former German standard DIN 1052 (Deutsches Institut für Normung [DIN], 2008); the implementation in Eurocode 5, transforms it into mean densities. For steel-to-timber or concrete-to-timber connections, the stiffness K_{ser} should be based on the density of the timber member and may be multiplied by 2.0. In other words, the slip $\delta_{steel} = 2\,\delta_{wood}$.

Lemaître et al. (2018) compared the EC5 approach with their advanced model and observed that the former underestimates the impact of dowel diameter. They showed that the slenderness of the fastener impacts the slip modulus, with lower values for the slenderness indicating failure modes without plastic hinges in the fastener.

4.3.2 Swiss Standard SIA 265

In the Swiss standard for the design of timber structures SIA 265 (SIA, 2012) the slip modulus K_{ser} in the serviceability limit state is defined as the ratio between the load F_y and the deformation v_y at the yield point. The standard gives formulas to determine K_{ser} for connections with different fasteners. For nails without predrilling in connections for service class 1 under short-term loading, loaded parallel-to-grain stiffness can be estimated as $K_{ser} = 60d^{1.7}$. For dowels, bolts and predrilled nails, it proposes the relationship $K_{ser} = 3\,\rho_k^{0.5} d^{1.7}$ for parallel-to-grain loading. For connections with fasteners loaded perpendicular to the grain, it recommends reducing the respective values by 50%. For the derivation of the equations for the slip modulus in SIA 265 (SIA, 2012), see Dubas (1981).

In general, the slip modulus *K* for connections without predrilling can be expressed as $K = C \cdot d^{exp}$, where *C* is a constant and *exp* is an exponent value. The SIA Standard used values of C = 60 and exp = 1.7. When performing the fitting with the data used by Dubas (1981), a constant C = 46.9 and an exponent exp = 1.87 were obtained and used in the standard (Jockwer et al., 2021).

4.3.3 New Zealand/Australian Standard NZS AS 1720.1

The updated New Zealand Standard 1720 (Standards New Zealand, 2022) has a section on deformations in various connections. For nailed connections, it suggests that the deformation can be approximated using the following four choices:

(a) A load equal to the short-term design strength of a single nail gives an average deformation δ = 2.5 mm;

(b) From 0 to 0.5 mm, one can calculate the average deformation using the following formula:

$$\delta = \frac{0.8 \cdot k_{37} \cdot N^{*2}}{\left(n_{\alpha,y,u}\right)^2} \tag{44}$$

where

k_{37} is the duration of load fastener deformation factor, which can be taken as 1.0 for loads acting for less than 5 minutes;

N^* is the specified nail load, in N; and

$n_{\alpha,y,u}$ is ultimate yielding strength of the nail, in N;

(c) From 0.5 to 2.5 mm, one can determine the deformation by interpolating the values obtained from (a) and (b);

(d) Above 2.5 mm the deformation, the load could increase 20% to 40% to give a maximum load at a slip between 6 and 10 mm.

The standard also provides an equation for calculating the average lateral deformation Δ_l of connections with bolts and dowels:

$$\Delta_l = \frac{N_a\, K_{37}}{n\, K_{ser}} \tag{45}$$

where

N_a is the specified load applied to the bolted or doweled connection, in N;

N is the number of fasteners in the connection; and

K_{ser} is the stiffness of the bolt or dowel connection.

The calculation of this last value is as follows:

$$K_{ser} = \frac{\rho'^{\,1.5}\, D}{12.6} \tag{46}$$

where $\rho'^{1.5}$ is the characteristic density in kg/m^3 and D is the fastener diameter in mm.

The New Zealand standard also provides an equation for calculating the average lateral deformation Δ_l of connections with coach screws (usually referred to as lag screws in Canada):

$$\Delta_l = \frac{N_a \, K_{37}}{n \, D \, t_2 \, K_{ser,\alpha}} \tag{47}$$

where

N_a is the specified load applied to the connection, in N;

n is the number of fasteners in the connection;

t_2 is the screw penetration length; and

$K_{ser,\alpha}$ is the lateral stiffness of the screw at an angle of α.

For loads parallel to the grain, the stiffness $K_{ser,0}$ is as follows:

$$K_{ser,0} = (0.0075\rho' - 0.29)k_{sp}k_{15} \tag{48}$$

where k_{sp} is the side plate factor (1.0 for timber and 1.5 for steel) and k_{15} is the timber seasoning factor (1.0 for unseasoned timber and 1.25 for seasoned timber, both loaded parallel to the grain). For loading perpendicular to grain, the factor $K_{ser,90}$ is

$$K_{ser,90} = (0.0075\rho' - 0.29)k_{90}k_{15} \tag{49}$$

where k_{90} is a factor for perpendicular-to-grain loading, $k_{90} = 2.12\, D^{-0.45}$.

4.3.4 Canadian Standard for Engineering Design in Wood CSA 086

The Canadian Standard for Engineering Design in Wood first included a formula for the lateral slip resistance of lag screws in the 1984 edition of CSA O86. More recent editions moved this formula to the Appendices, so according to the current Clause A.12.6.5.4, the lateral deformation of lag screw connections is

$$\Delta = \frac{P}{k \, d_F t_2 n_F} \tag{50}$$

where

Δ is the lateral deformation, in mm;

P is the specified load on the connection, in N;

k is the lateral slip resistance of a lag screw, in MPa per mm of slip;

d_F is the diameter of the lag screw;

t_2 is the length of penetration into the wood member; and

n_F is the number of fasteners in the connection.

The calculation of the lateral slip resistance *k* uses the below equations for parallel-, perpendicular-, and angle-to-grain loading, respectively:

$k = k_p = (5.04\ G - 0.29)\ J_y\ J_G\ K_{SF}$ for parallel-to-grain loading ($\theta = 0°$);

$k = k_Q = (5.04\ G - 0.29)\ J_Q\ J_G\ K_{SF}$ for perpendicular-to-grain loading ($\theta = 90°$); and

$k = k_p\ k_Q / (k_p \sin^2\theta + k_Q \cos^2\theta)$ for angle-to-grain loading ($0° < \theta < 90°$),

where

- G is mean relative density;
- J_Y is side plate material factor, which is 1.15 for steel plates;
- J_G is group-action factor;
- J_Q is perpendicular-to-grain load factor, ranging from 0.5 for a 1 in. (25.4 mm)–diameter lag screw to 1.0 for a 3/16 in. (4.7mm)–lag screw; and
- K_{SF} is service condition factor.

Clause A12.9.3.3 of CSAO86 provides a lateral deflection calculation for wood-to-wood connections with nails, spikes, or wood screws, including structural panel-to-lumber connections. It states that for specified loads, P, not greater than $n_u/3$, the lateral deformation of these connections is as follows:

$$\Delta = 0.5\ d_F\ K_m \left(\frac{P}{n_u}\right)^{1.7} \tag{51}$$

where

- Δ is lateral deformation, in mm;
- d_F is nominal diameter of fastener, in mm;
- K_m is service creep factor, varying from 1.0 to 3.0;
- P is specified load per fastener per shear plane, in N; and
- N_u is unit lateral strength resistance per shear plane, in N.

5 PERFORMANCE OF CONNECTIONS IN MASS TIMBER DIAPHRAGMS

This section provides experimental and analytical research on the connections used in mass timber diaphragms. The material presented here will provide designers with important information on the values for strength, stiffness, and ductility of certain connection configurations used in mass timber diaphragms.

5.1 Half-Lap and Spline Connections Using STSs in CLT

Gavric et al. (2015) conducted an extensive experimental programme on typical CLT connections with STSs. Their study performed in-plane monotonic and cyclic shear, as well as withdrawal, tests on wall-to-wall, floor-to-floor, and wall-to-floor CLT connections in 12 different configurations. Some tested configurations that apply to diaphragm applications are shown in Figure 34. Configurations #1 to #4, although designated as wall connections in the study, equally apply to diaphragms, while configurations #11 and #12 apply to diaphragm applications only. The authors evaluated mechanical properties such as strength, stiffness, energy dissipation, ductility, and strength degradation for all configurations.

Test configuration	Loading direction	Type and number of screws
Wall–Wall panel connections (parallel panels)		
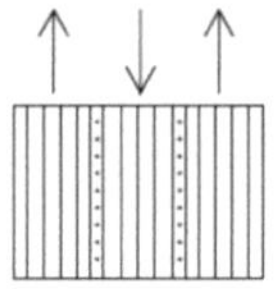 Test #1	Lateral parallel (lap joint)	HBS Φ8x80mm 2x2 screws
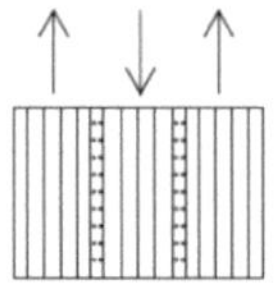 Test #2	Lateral parallel (spline joint)	HBS Φ8x80mm 2x4 screws
 Test #3	Lateral perpendicular (lap joint)	HBS Φ8x80 2 screws
 Test #4	Lateral perpendicular (spline joint)	HBS Φ8x80 2x2 screws

Test configuration	Loading direction	Type and number of screws
Floor–Floor panel connections		
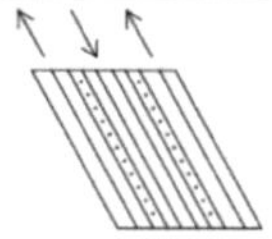 Test #11	Lateral parallel (lap joint)	HBS Φ10x140 2x2 screws
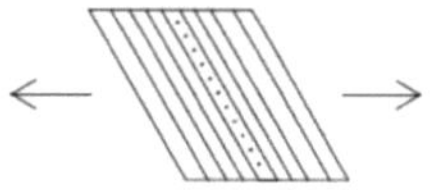 Test #12	Lateral perpendicular (lap joint)	HBS Φ10x140 2 screws

Figure 34. Test matrix for CLT connections with STSs that apply to diaphragm situations (Gavric et al., 2015).

Connection configurations #1 to #4 used 5-layered CLT where each layer is 17 mm thick, for a total panel thickness of 85 mm. Panels in test configurations #11 and #12 had a 6-layered build-up structure with a double layer in one direction in the middle (27-17-27-27-17-27), forming a panel with a total thickness of 142 mm. Connections had two different types of joints: half-lap and spline joints. Half-lap joints had an overlap length of 50 mm in configuration #1 and of 120 mm in configuration #11. Spline joints used a Kerto LVL strip 28 mm thick and 180 mm wide. In test configurations #1 and #11, panels directly connected to each other, with two STSs on each side spaced 10 cm apart. Test configuration #2 used two rows of screws to connect the LVL strip to the panels, also with 10 cm spacing in vertical direction. Similarly, test configurations #3, #4, and #12 had screws spaced at 10 cm. The overlap dimensions and screw types in test configuration #3 were the same as in test configuration #1. Test configuration #4 was similar to test configuration #2, while test configuration #12 had the same connection detail as test configuration #11.

Experimental results were assessed in terms of strength, stiffness, energy dissipation, and ductility, following the standard procedure given in EN 12512 (2001). Six replicates from each connection configuration were tested. The values of mechanical properties for each specimen were analyzed by taking into account results from both sides of hysteretic loops, except for the cases loaded in tension only when that was the only available direction (Configurations #3, #4, and #12). The test results are summarized in the table in Figure 35. All results are presented per one screw, except for Test #2 and Test #4 where results are presented for two screws as the spline connection needs at least two fasteners to provide viable connection between the two panels.

Mechanical property	Test configuration											
	1		2		3		4		11		12	
	x_{mean}	COV	x_{mean}	COV	x_{mean}	COV	x_{mean}	COV	x_{mean}	COV	x_{mean}	COV
k_{el} (kN/mm)	1.24	18.59	0.84	16.28	1.25	29.65	0.94	17.02	1.27	20.67	0.83	6.98
k_{pl} (kN/mm)	0.11	20.01	0.10	39.81	0.21	29.65	0.16	17.02	0.21	20.67	0.14	6.98
F_y (kN)	3.23	11.95	4.85	17.67	2.91	17.64	3.23	7.02	6.67	10.23	6.91	9.36
v_y (mm)	2.55	11.99	5.70	22.75	2.34	24.17	3.13	20.05	5.39	26.91	7.30	14.78
F_{max} (kN)	5.25	13.18	7.33	11.89	4.91	24.41	6.40	4.96	10.10	4.28	9.83	11.75
$F_{max(3rd)}$ (kN)	3.69	14.93	5.24	12.61	4.07	23.92	5.62	5.77	7.39	8.80	7.75	22.93
v_{max} (mm)	23.50	9.81	34.37	17.40	17.51	28.84	39.40	8.90	29.83	0.12	31.59	28.22
F_u (kN)	4.20	13.18	5.86	11.89	3.93	24.42	5.12	4.94	8.08	4.29	7.87	11.75
v_u (mm)	31.55	2.48	37.66	8.79	23.76	17.79	50.52	16.28	33.59	18.66	41.17	15.61
D (–)	12.81	10.34	7.15	21.20	10.71	31.69	16.36	14.18	6.58	10.23	5.74	23.85
D_{mon} (–)	15.98	–	14.43	–	20.97	–	15.66	–	13.44	–	8.99	–
F_{30} (kN)	4.61	15.98	7.81	12.59	–	–	6.02	2.82	9.44	13.31	9.16	15.78
D_{30} (–)	12.12	11.01	7.30	21.49	–	–	9.91	20.19	6.14	21.73	4.20	16.87
$\Delta F_{1\text{-}3}$ (%)	29.61	17.67	28.37	16.12	13.93	14.10	10.44	5.73	26.73	24.54	13.17	13.04
$v_{eq(1st)}$ (%)	14.52	6.71	14.72	10.48	5.98	10.56	5.81	9.41	22.20	4.12	8.58	13.73
$v_{eq(3rd)}$ (%)	9.12	23.37	11.27	14.86	2.97	19.30	3.30	13.93	15.17	7.63	2.49	12.16
$F_{0.05}$ (kN)	3.81		5.62		2.64		5.23		9.13		7.30	
$F_{0.95}$ (kN)	7.14		8.56		8.67		7.45		11.16		13.08	
γ_{Rd} (–)	1.88		1.52		3.28		1.42		1.22		1.79	

Figure 35. Test results from six tested configurations of CLT connections with STSs (Gavric et al., 2015).

The mechanical properties presented in Figure 35, derived from EN12512, are as follows: K_{el} and K_{pl} are the initial and plastic stiffness, respectively; F_y and v_y are yielding load and yielding displacement, respectively; F_{max} and v_{max} denote maximum load and maximum displacement, respectively; F_u and v_u represent the ultimate load and ultimate displacement, respectively; $F_{max(3rd)}$ is the strength of the third backbone curve; Δ_{F1-3} is the percentage difference between first and third backbone strength; D is the ductility (ratio between ultimate displacement and yield displacement) from cyclic tests; D_{mon} is the ductility ratio from monotonic tests. As one of the criteria for ultimate values in EN 12512 is the strength at 30 mm displacement, the table also includes force F_{30} and ductility ratio D_{30}. it also presents energy dissipation properties in terms of the equivalent viscous damping for the first and third cycles of the hysteretic loop $v_{eq(1st)}$ and $v_{eq(3rd)}$, respectively. Finally, $F_{0.05}$ and $F_{0.95}$ are the 5th and 95th percentiles of the resistance, respectively, while γ_{rd} is the overstrength factor.

The half-lap joints from configuration #1 exhibited 50% higher initial stiffness than the spline joints (Config. #2). The connections in configuration #2 had 40% higher resistance, which was reached at a 46% higher displacement level than for the connections in configuration #1. The ultimate displacement of the spline joint (Config. # 2) was higher, with an average value of 37.7 mm as compared to 31.4 mm for the half-lap joint (Config. #1). The screws in half-lap joints started to fracture after reaching relatively high displacement levels, something the authors did not observe in connections with LVL spline joints. The LVL spline joint had lower ductility because the test ended upon reaching the 30 mm slip, as required by EN12512. If the test had continued, the ultimate slip would have probably doubled, leading to roughly the same ductility. In terms of energy dissipation capacity, both types of connections exhibited similar damping ratios in the third cycle.

Connection configuration #11, with a thicker CLT panel, longer screws with larger diameters, and a longer overlap, exhibited higher yield load, yield displacement, and average strength, but lower ductility than configuration #1. The damping ratio at the third cycle was 66% higher, as well. The failure modes for both half-lap and spline connections were "Mode E," according to CSA O86 and Eurocode 5, with the screws bending and one plastic hinge forming. While in some tests fracture of screws eventually occurred, no brittle failure modes were observed, and no splitting occurred during the monotonic and cyclic tests.

When the tests were performed in the perpendicular (tension) direction (Configurations #3, #4, and #12), the failure modes for half-lap connections were different than for those with a LVL spline. Half-lap connections (configurations #3 and #12) exhibited a failure mode that was generally due to either the panel splitting of the panel or the bonds failing in the inner layers. Plug shear in the zone of the screw thread was observed in some tests. On the other hand, for the LVL spline connection (configuration #4), the observed failure mode was mainly associated with pull-through of the screw head, forming one plastic hinge within the CLT panel. A comparison of the results from test configurations #3 and #4 showed a significant difference in the initial stiffness. Half-lap joints exhibited 33% higher initial stiffness than LVL splines. Spline joints, however, could resist higher forces at displacements that were more than twice as large. This was mainly because of the different type of failure mode: brittle failure modes in half-lap joints prevented the generation of higher resistances. Connection configuration #12 (120 mm lap with a 142 mm–thick CLT panel) performed better than configuration #3 (50 mm lap with a 85 mm–thick CLT panel), as the authors observed no brittle failure modes. This configuration reached ultimate values at relatively large displacements, on average more than 40 mm. The ductility ratio was relatively low, mainly due to the higher yielding displacement. The achievement of ductile failure modes in CLT connections, which is crucial for seismic design, requires one to use the end and edge distance requirements reported by Uibel and Blaß (2006, 2007). Spacing for STSs in the draft CSA O86 provisions for STSs will be sufficient to ensure ductile failure modes.

The overstrength factor (designated as γ_{Rd} in Gavric et al., 2015) is defined as the ratio between the 95th percentile of the connection strength distribution and the analytical prediction of the design connection strength F_d. Since current regulations such as Eurocode 5 (CEN, 2004a) do not provide analytical design formulas for the prediction of the load-carrying capacities of connections with STSs in CLT members, the authors calculated the design value F_d by dividing the characteristic experimental strength $F_{0.05}$ (the 5th percentile value) by the strength partial factor $\gamma_M = 1.0$, as per Eurocode 8 (CEN, 2004b) for dissipative timber structures. The average overstrength factor for tests with ductile failure modes was 1.56. If the connection requirements to achieve ductile failure modes are satisfied, it is possible to use a conservative overstrength factor $\gamma_{Rd} = 1.6$ for CLT connections with STSs. This factor may be even lower if the number of tested specimens is larger and if there are more screws in a row, as the scatter of the strength decreases when more screws "work" simultaneously.

Finally, the authors compared the mean slip modulus of the connections with values obtained using the analytical equations currently included in Eurocode 5 for dowel connections. EC5 empirical formulas for the prediction of the screw connection slip modulus at the serviceability limit state corresponded well with the experimental elastic values.

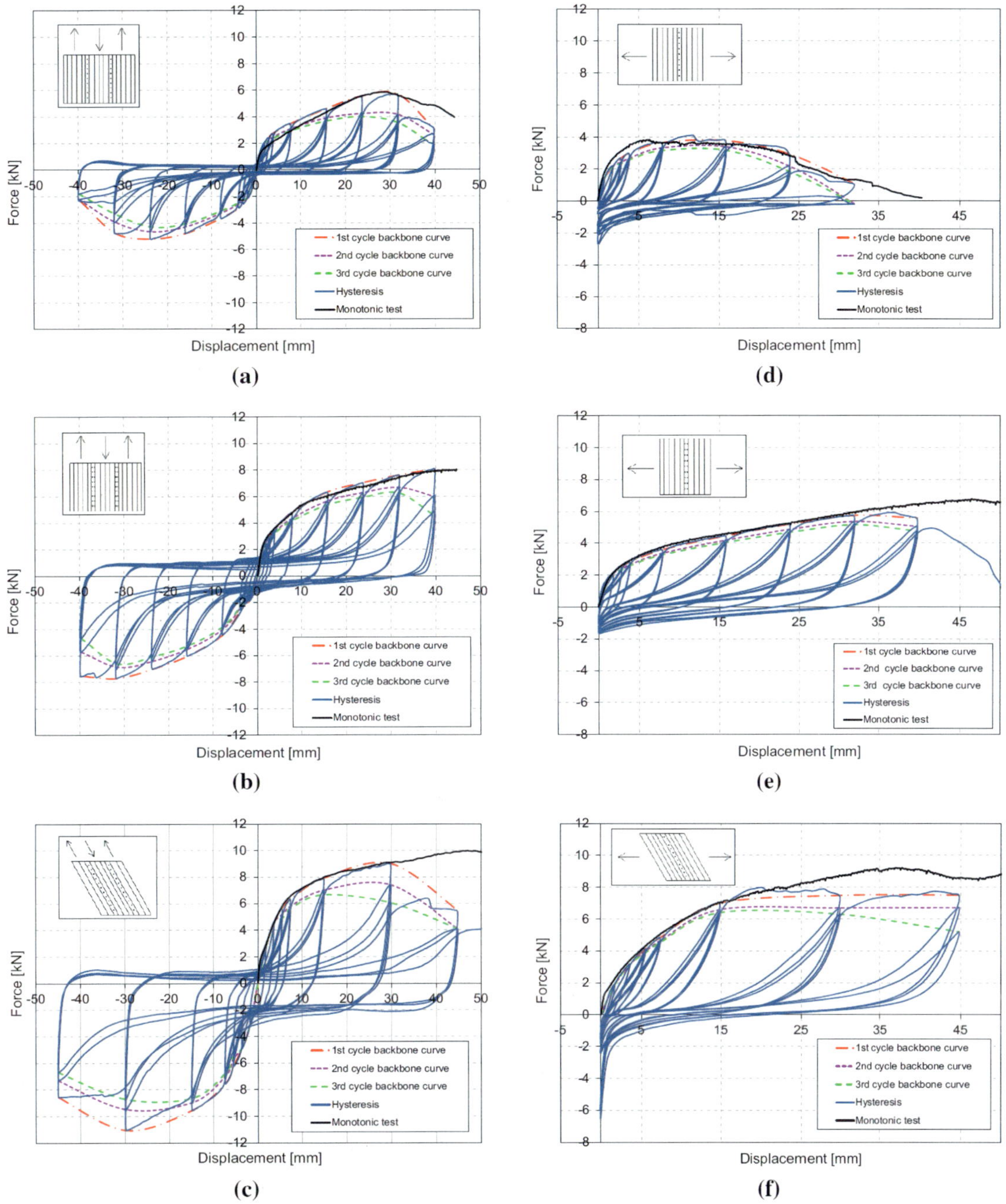

Figure 36. Typical hysteresis loops from cyclic tests along with the backbone curves and monotonic curves for different connection configurations: (a) Config. #1; (b) Config. #2; (c) Config. #11; (d) Config. #3; (e) Config. #4; (f) Config. #12 (Gavric et al., 2015).

Figure 36 shows the hysteresis loops of the tested connections with first, second, and third cycle backbone curves, along with the monotonic curve. Comparing the monotonic curve with the first backbone curve shows little difference between them, particularly before reaching the failure loads. The average degradation of peak strength from the first backbone curve to the third ranged from 10.4% to 29.6%, depending on the type of specimen.

5.2 Connections in CLT Using Double Inclined STSs

Hossain et al. (2016) investigated the shear resistance and performance of STSs in three-ply CLT panels. Specifically, they looked at the feasibility of using STSs with double inclined fasteners for shear connections in CLT panels that use butt joints. The data obtained from these tests allow engineers to specify an innovative connection assembly that utilizes double inclined STSs in butt-jointed CLT panels that may be part of a mass timber diaphragm.

The test specimens consisted of assemblies of three CLT panels (99 mm thick, 500 mm wide, and 1,500 mm long), exhibiting two shear planes. The CLT panels consisted of three layers, with the outer two being 32 mm thick and the middle one being 35 mm thick. The specimens used fully threaded STSs (ASSY VG CYL, SWG Schraubenwerk Gaisbach, Waldenburg, Germany), 8 x 180 mm. The STSs were installed at an angle of 45° to the joint line between the CLT panels and at an angle of 32.5° to the faces of the panels (Figure 37). This double inclined screw installation resulted in a compound angle α between the grain direction of the wood and the screw axis of 53.4°, with the screws being loaded in withdrawal. Because the grain direction of the outer layers was parallel to the shear load, the angle between the load and the screw axis (subsequently labelled β) was also 53.4°.

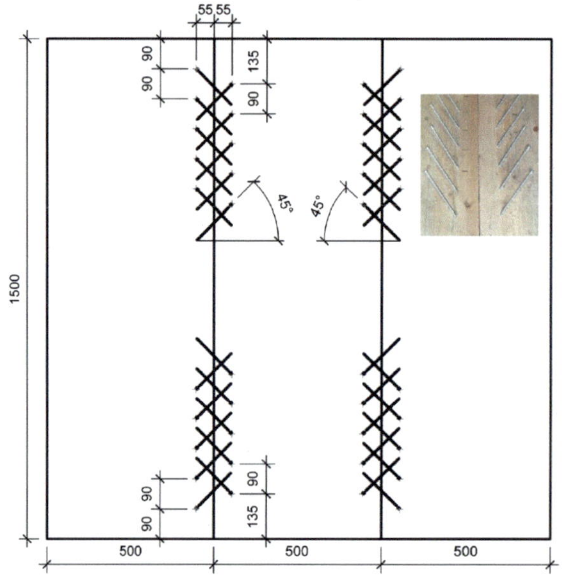

Figure 37. Screw assembly layout for test specimen, with photo of detail (Hossain et al., 2016).

All STSs were installed from the same panel face, but with four different orientations at each shear plane (Figure 37), so that half of the screws were in tension and half were in compression, in each direction of the shear loading. The STSs were spaced at 90 mm on centre, exceeding the minimum spacing requirements of 5d (ETA-11/0190). They were installed with an end distance of 90 mm, and the rows were offset by 45 mm to avoid interference between the STSs. The distance in the centre of the panel between the top and bottom groups of STSs was 600 mm. The screws were installed in 4.8-mm diameter predrilled holes.

The experimental program, consisting of seven quasi-static loading tests (S1–S7) and four cyclic tests (C1–C4), took place at UBC's Structures Laboratory.

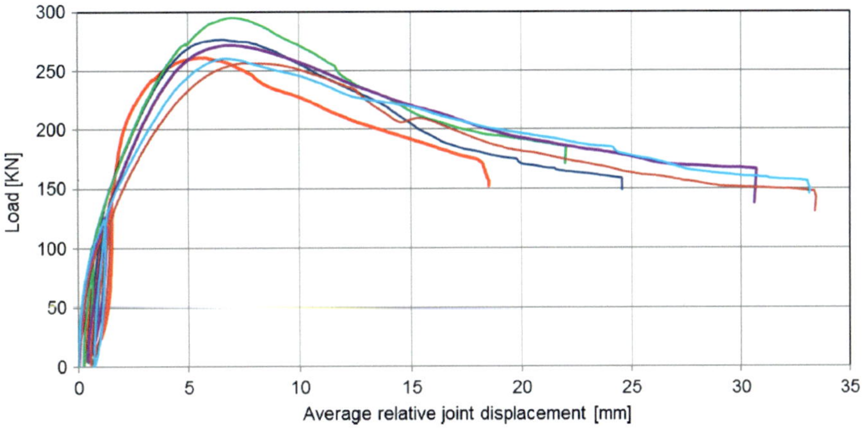

Property	S2	S3	S4	S5	S6	S7	Mean	COV (%)
Capacity (kN/screw)	6.5	6.9	7.4	6.8	6.5	6.5	6.8	5
EEEP yield (kN/screw)	6.0	6.2	6.5	6.0	5.6	5.7	6.0	5
Display at EEEP (mm)	2.5	1.7	1.9	2.0	2.0	1.6	2.0	16
Display at failure (mm)	12.3	13.6	12.5	15.4	16.2	17.1	14.5	14
Stiffness (kN/mm/screw)	4.0	5.8	5.5	5.0	4.7	5.7	5.1	14
Ductility ratio (μ)	4.9	8.0	6.6	7.7	8.1	10.7	7.7	25

Figure 38. Load-displacement curves from static tests (top); Connection properties from monotonic tests (bottom) (Hossain et al., 2016).

The load-deformation curves of samples S2–S7, as obtained from the static tests, appear in Figure 38 top. The plotted displacement is the average of the two linear transducer measurements at each shear plane. Figure 38 bottom provides a summary of the results from the static testing. The ASTM E2126 equivalent energy elastic-plastic (EEEP) procedure served to determine the properties. The maximum load of the connections averaged around 90 kN/m, while the yield loads were 80 kN/m (6.8 and 6.0 kN/screw, respectively). The differences between results from the test specimens were relatively small, with a coefficient of variation (COV) of 5%. The displacements at yield and at failure averaged 2.0 and 14.5 mm, respectively, with larger COVs of 16% and 14%, respectively. In accordance with EN29816 (CEN, 1991), the computation of the average connection stiffness was based on a loading range between 10% and 40% as 5.1 kN/mm/screw. Finally, the connection ductility μ, defined as the ratio between the displacement at failure and the yield displacement determined using the EEEP method, was 7.7 on average for the monotonic tests.

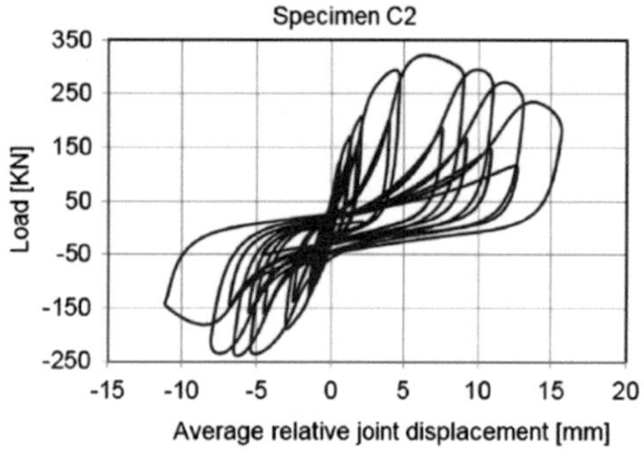

Property	C1	C2	C3	C4	Mean	COV (%)
Capacity (kN/screw)	6.5	7.1	6.6	6.5	6.7	4
Capacity[a] (kN/screw)	5.6	6.0	5.5	5.6	5.7	4
EEEP yield (kN/screw)	5.9	6.4	5.8	5.7	5.9	5
EEEP yield[a] (kN/screw)	5.3	5.5	4.7	4.9	5.1	7
Displacement at EEEP (mm)	2.1	2.1	1.5	1.8	1.9	15
Displacement at failure (mm)	7.8	7.8	7.4	7.2	7.5	4
Ductility ratio (μ)	3.7	3.7	4.9	4.0	4.1	14
Damping, $\xi_{eq(1st)}$ (%)	6.7	7.4	6.7	6.5	A6.8	6
Damping, $\xi_{eq(3rd)}$ (%)	16.5	14.1	15.4	15.3	15.3	7

[a]Denotes values from weaker envelope only.

Figure 39. Typical cyclic behaviour obtained from the cyclic tests (left); Connection properties from cyclic tests (right) (Hossain et al., 2016).

The results from the cyclic tests appear in Figure 39 right. The EEEP calculations were done for the first envelope, following the same procedure as for the static tests. The capacities and yield loads averaged 89 and 79 kN/m (6.7 and 5.9 kN/screw), respectively. As a conservative measure, Figure 39 right also presents the ultimate and EEEP yield loads for the weaker (third) envelope. The weaker envelope capacities and yield loads were approximately 15% smaller and averaged 76 and 68 kN/m (5.7 and 5.1 kN/screw), respectively. Like the static tests, the differences between test specimens were relatively small, with COVs ranging from 4% to 7%. The displacements at yield and at failure averaged 1.9 and 7.5 mm, respectively. The average ductility ratio in the cyclic tests was 4.1.

5.3 Connections in CLT with STSs in Shear and Withdrawals

Hossain et al. (2018) investigated a novel connection assembly combining STSs that act in shear and withdrawal to optimize the strength and deformation properties of the connections. The data obtained from this study can help engineers specify connection assemblies for LLRSs, specifically shear walls and floor diaphragms of CLT structures.

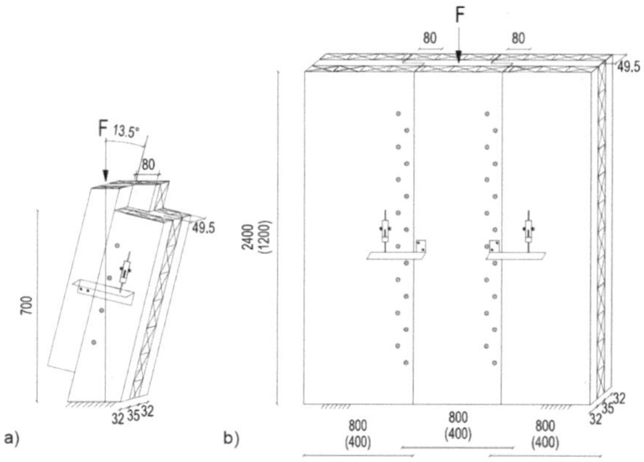

Series	CLT panel size	STS action	Loading	STS per shear plane	Replicates
S-S-M	Small	Shear	Monotonic	4	6
S-W-M	(290 × 700 mm) one	Withdr.	Monotonic	6	6
S-C-M	shear plane	Combined	Monotonic	8	6
M-S-M	Medium	Shear	Monotonic	16	1
M-S-C	(400 × 1200 mm) two		Cyclic		3
M-W-M	shear planes	Withdr.	Monotonic		1
M-W-C			Cyclic		3
M-C-M		Combined	Monotonic		1
M-C-C			Cyclic		3
L-S-M	Large	Shear	Monotonic	32	1
L-S-C	(800 × 2400 mm) two		Cyclic		2
L-W-M	shear planes	Withdr.	Monotonic		1
L-W-C			Cyclic		2
L-C-M		Combined	Monotonic		1
L-C-C			Cyclic		2

Figure 40. Test specimen layout and testing series: (a) small specimens with one shear plane; (b) medium and large specimens with two shear planes; test series overview (right). (Hossain et al., 2018).

Connection specimens consisted of 3-ply, 99 mm–thick CLT panels with half-lap joints. The laps were 80 mm wide and 50 mm deep (half the CLT panel thickness). Three different specimen sizes were fabricated and tested: Small, Medium, and Large. In small specimens, two individual panels, each 290 mm wide and 700 mm long, were connected using one shear plane (Figure 40a). Small specimens, depending on the screw action, involved four, six, or eight STSs (Figure 40 right). Six replicates were tested under monotonic loading. Medium size specimens had three individual panels connected in two shear planes. Each panel was 1200 mm long and 400 mm wide (Figure 40b) and had 16 STSs installed in each shear plane. One monotonic and three cyclic replicates of each series were tested. Large size specimens also had three individual panels connected in two shear planes. Each panel was 2400 mm long and 800 mm wide (Figure 40), with 32 STSs installed in each shear plane. One replicate under monotonic loading and two replicates under reversed cyclic loading were tested for each series. All assemblies were connected with STSs designed for three different actions (Figure 41): (1) Shear action, with STSs placed at 90° to the load; (2) Withdrawal action, with STSs placed at 45° to the load; and (3) Combined action, with STSs placed at both 90° and 45° to the load. For an overview of all test series, see Figure 40 right. Test series labelling combined the specimen size, the STS action, and the loading type, e.g., test series *S-S-M* stands for a test series of small panels, connected by STSs loaded in shear under monotonic loading.

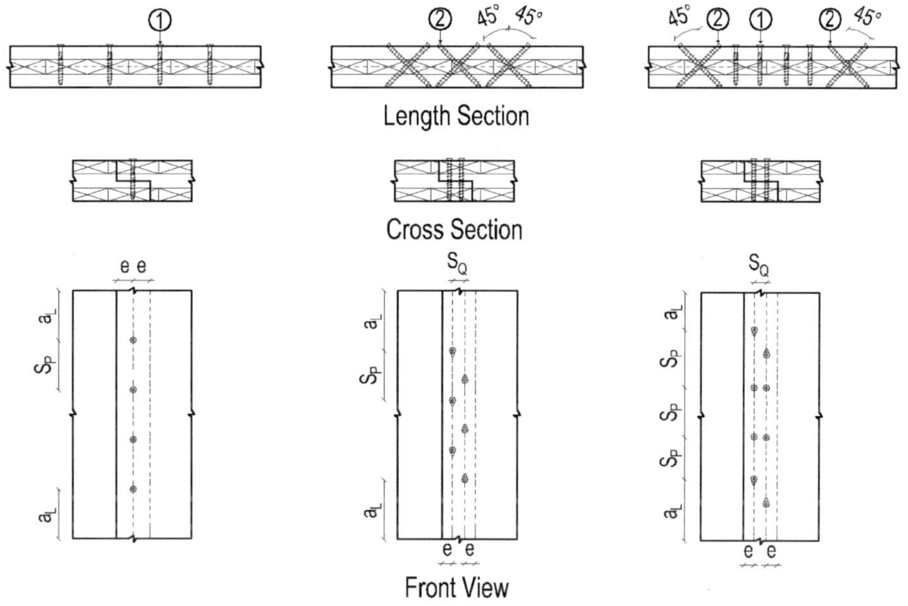

Figure 41. Screw layout: shear (left); withdrawal (centre); combined (right). Fastener 1: ASSY 3.0 ECOFAST Screw PT 8 mm × 90 mm; fastener 2: ASSY Plus VG Screw FT 8 mm × 140 mm (Hossain et al., 2018).

The STSs had two different orientations at each shear plane, so that half of the screws were in tension and half were in compression in each direction of the shear loading. The predrilled holes extended through only one panel using a 4.8 mm diameter drill bit, and this partial predrilling of a pilot hole should not have affected the connection performance. Specimens that used STSs loaded either only in shear or only in withdrawal had spacing parallel to grain of S_P = 120 mm (15 d), loaded end distance a_L = 120 mm (15 d), spacing perpendicular to grain S_Q = 32 mm ($\geqslant$ 3 d) for fully threaded screws, and unloaded edge distance e = 40 mm (5 d) for partially threaded screws and e = 24 mm (3 d) for fully threaded screws (Figure 41a and b). In the specimens combining

STSs loaded in shear with STSs loaded in withdrawal (Figure 41c), the unloaded edge distance *e* = 3 d was below the requirements (5 d) for partially threaded screws. This reduced spacing was a consequence of keeping the lap width of 80 mm constant for all test specimens. Since no splitting occurred during installation or testing, this reduced spacing did not impact the results.

All CLT panels were the products of a Canadian manufacturer following the ANSI/APA PRG-320 standard. The wood species combination was SPF (Spruce-Pine-Fir), with a mean oven-dry relative density of 0.42; the adhesive was Purbond Polyurethane. The CLT grade was V2M1, where the parallel layers are SPF No. 1/No. 2 and perpendicular layers are SPF No. 3. The average moisture content of the CLT at the time of connection testing was 10% (+/−2%).

Series	F_{max}[a] [kN]	F_Y[a] [kN]	d_{Fmax} [mm]	d_{FY} [mm]	d_{ult} [mm]	μ [–]	k[a] [kN/mm]
S-S-M	6.4 (5.7)	5.6 (16)	25 (34)	6.8 (54)	34 (33)	4.0 (34)	1.9 (23)
S-W-M	7.6 (9.5)	7.1 (8.1)	2.7 (19)	1.8 (29)	6.7 (17)	1.6 (41)	9.4 (61)
S-C-M	6.5 (7.7)	5.9 (8.4)	16 (59)	1.0 (30)	18 (72)	18 (7.5)	12 (48)
M-S-M[b]	7.3 (–)	6.2 (–)	31 (0)	6.7 (38)	42 (38)	5.0 (20)	1.8 (6.7)
M-S-C	5.5 (6.7)	4.7 (6.4)	23 (18)	5.1 (39)	36 (30)	4.2 (72)	0.9 (5.3)
M-W-M[b]	7.2 (–)	6.7 (–)	2.8 (55)	2.0 (41)	6.5 (16)	1.4 (26)	8.6 (69)
M-W-C	6.5 (12)	6.3 (5.3)	2.5 (37)	1.4 (44)	7.2 (55)	2.0 (78)	6.8 (69)
M-C-M[b]	5.5 (–)	5.0 (–)	31 (4.6)	1.5 (88)	43 (86)	34 (39)	8.2 (3.3)
M-C-C	5.4 (5.7)	5.1 (5.0)	15 (44)	1.3 (56)	26 (73)	15 (41)	4.3 (32)
L-S-M[b]	6.2 (–)	5.8 (–)	33 (2.2)	5.4 (3.8)	41 (6.0)	6.0 (24)	2.0 (14)
L-S-C	5.8 (18)	5.0 (19)	21 (37)	5.7 (51)	29 (72)	4.9 (45)	1.1 (22)
L-W-M[b]	6.5 (–)	6.0 (–)	3.6 (22)	1.9 (16)	7.2 (6.3)	1.9 (18)	6.5 (19)
L-W-C	5.2 (10)	4.9 (15)	1.9 (41)	1.4 (32)	2.0 (9.8)	1.3 (23)	4.2 (38)
L-C-M[b]	5.0 (–)	4.5 (–)	15 (10)	1.3 (21)	42 (21)	12 (5.1)	7.5 (22)
L-C-C	4.9 (12)	4.3 (13)	6.0 (39)	1.2 (44)	23 (90)	7.0 (38)	4.7 (14)

[a] Values presented per single screw.
[b] Average deformation, stiffness and ductility were calculated based on the measurement of the two shear planes of the single test specimen.

Figure 42. Summary of the averages of the test results, with COV in parenthesis (Hossain et al., 2018).

Results from the testing (Figure 42) showed that using STSs in shear leads to ductile connections ($\mu > 4$) that can reach large displacements (> 30 mm) before failure. Assemblies with STSs loaded in shear are appropriate for applications requiring connection ductility and deformation capacity. Using STSs in withdrawal leads to significantly stiffer joints (on average, stiffness increases fourfold) than when using STSs in shear. These joints, however, are brittle and fail at very small displacements (< 2 mm); they should not be used as energy-dissipative connections for seismic applications. The combination of STSs in withdrawal and shear allows joints to exhibit both high stiffness and high ductility. The stiffness of such joints was very similar to the stiffness of joints using STSs in withdrawal only, while the ductility was similar to that of joints using STSs in shear only. The performance of shear connections under cyclic loading was found to slightly differ from that of monotonic loading, with cyclic values reduced by anywhere between 11% (for load-carrying capacity) and 40% (for stiffness). This finding demonstrates the need to provide designers and engineers with data about connection performance under cyclic loading for reliable seismic performance. Finally, the load-carrying capacity and stiffness of connections with STSs loaded in withdrawal proved to be a function of the number of screws. In other words, a group effect was observed, i.e., a reduction of per screw load-carrying for larger specimens. The authors have since investigated this finding further and some of their findings appear in the next section.

5.4 Group Effect in CLT Connections Using STSs

The group effect in fasteners is well known, and as such, it has already been implemented in many national and international codes and standards. Reduction factors are included in material design standards to decease the strength per fastener when using multiple fasteners in a row parallel to grain. These reduction factors account for the fact that the load on each connector is not equal; the end connectors in a row parallel to the direction of the applied force have larger loads than those in the centre of the row. Thus, adding connectors to the row does not appreciably increase the strength of the connections, which is instead controlled by the end fasteners.

The U.S. AWC National Design Specification for Wood Construction (NDS) introduced the group-action factor C_g in 1973. Since 1997, the NDS has also provided a group-action factor for dowel-type fasteners, bolts, and lag screws with diameters between 6.4 and 25.4 mm (1/4 to 1 in.), split rings, and shear plates in a row, based on work by Zahn (1991). C_g is a function of the stiffness and cross-sectional area ratios of the two connected timber members, the number of fasteners, the spacing between them, and the load-slip modulus of the connection. C_g varies from 0.80 to 0.90, depending on the diameter size and the values of the other inputs. The New Zealand timber code (Standards New Zealand, 2022) tabulates the reduction factor for multiple bolts or "coach" screws (lag screws) in a connection. For 5 to 9 screws, the factor is 0.95; for 10 to 15 screws, the factor is 0.80; and for 16 or more screws, the factor is 0.62. CSA O86 introduced the modification factor J_G for bolts, lag screws, split rings, and shear plates in 1984. The year 1989 then brought adjustments to account for the number of bolts in a row, the loaded end distance, the number of bolt rows, and bolt slenderness, based on findings from Yasumura et al. (1987) and Masse et al. (1989). In 2009, CSA O86 introduced a mechanics-based model for the capacity of bolted and doweled connections that accounts for brittle failure modes and removed the group-effect modification factor for bolts. The provisions for lag screws, however, remained unchanged. Eurocode 5 (CEN, 2014a) provides different group effect factor equations for nails (a function of spacing and predrilling), bolts loaded laterally parallel to the grain (a function of diameter and spacing), and wood screws loaded in withdrawal (a function of the number of effective fasteners $n_{eff} = n^{0.9}$). The number of effective fasteners for bolts loaded perpendicular to grain, n_{eff}, is the same with the number of fasteners n, while n_{eff} remains unspecified in screwed connections loaded perpendicular to the grain.

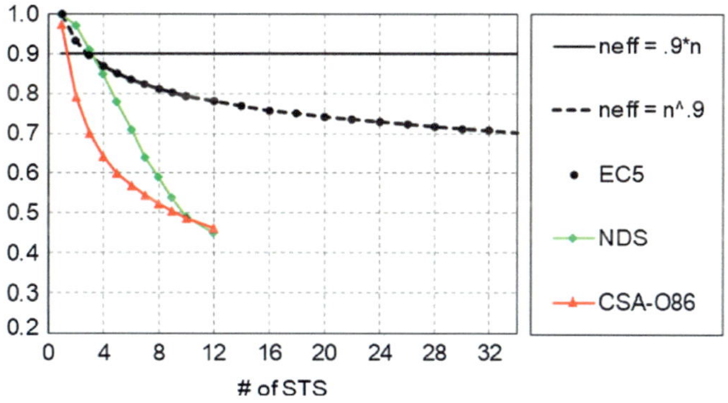

Figure 43. Reduction in strength for n_{eff}, according to different design standards and equations (Hossain et al., 2019).

With respect to the group effect in STSs, Brandner et al. (2016) investigated the withdrawal capacity of STS groups placed in the narrow face of CLT and conservatively proposed the relationship $n_{eff} = 0.9n$, even though test results indicated $n_{eff} > 1$. Krenn and Schickhofer (2009) performed tension tests on glulam joints with steel side plates using inclined STSs and showed that n_{eff} is variable and inconsistent when comparing ductile and withdrawal failure modes. They proposed $n_{eff} = 0.9n$ for the capacity reduction and $n_{eff} = n^{0.8}$ for serviceability limit states. They also studied the influence of number of screws on stiffness in steel-to-timber joints with inclined screws but proposed no equation for it. Ringhofer et al. (2018) pointed out that $n_{eff} = n$ is achievable if one avoids unfavourable brittle failure modes and if the screws and spacing between them provide enough ductility for load redistribution. Technical approvals for STSs, e.g., ETA-11/0190 (Deutsches Institut für Bautechnik [DIBt], 2018) provide $n_{eff} = \max(n^{0.9}, 0.9n)$ for axially loaded screws. The group-effect expressions implemented into Eurocode 5, the NDS, and CSA O86 differ significantly when applied to CLT shear connections using STSs, e.g., the value of n_{eff} according to Eurocode (EC5) is twice that of CSA O86 for a larger number of screws (Figure 43).

These discrepancies indicated the need for further fundamental research on multiple-screw timber connections. Hossain et al. (2019) investigated the group effect in CLT in-plane shear connections that utilize STSs, evaluating the influence of the number of STSs on the load-carrying capacity, stiffness, and ductility of connections under monotonic and cyclic loading for different connection types that utilize STSs acting in shear or withdrawal. This comprehensive study involved a total of 175 connections with spline and lap connections loaded in shear, butt joints loaded in shear and withdrawal, and lap joints loaded in withdrawal. The number of STSs in a row for the tested connections varied between 2 and 32.

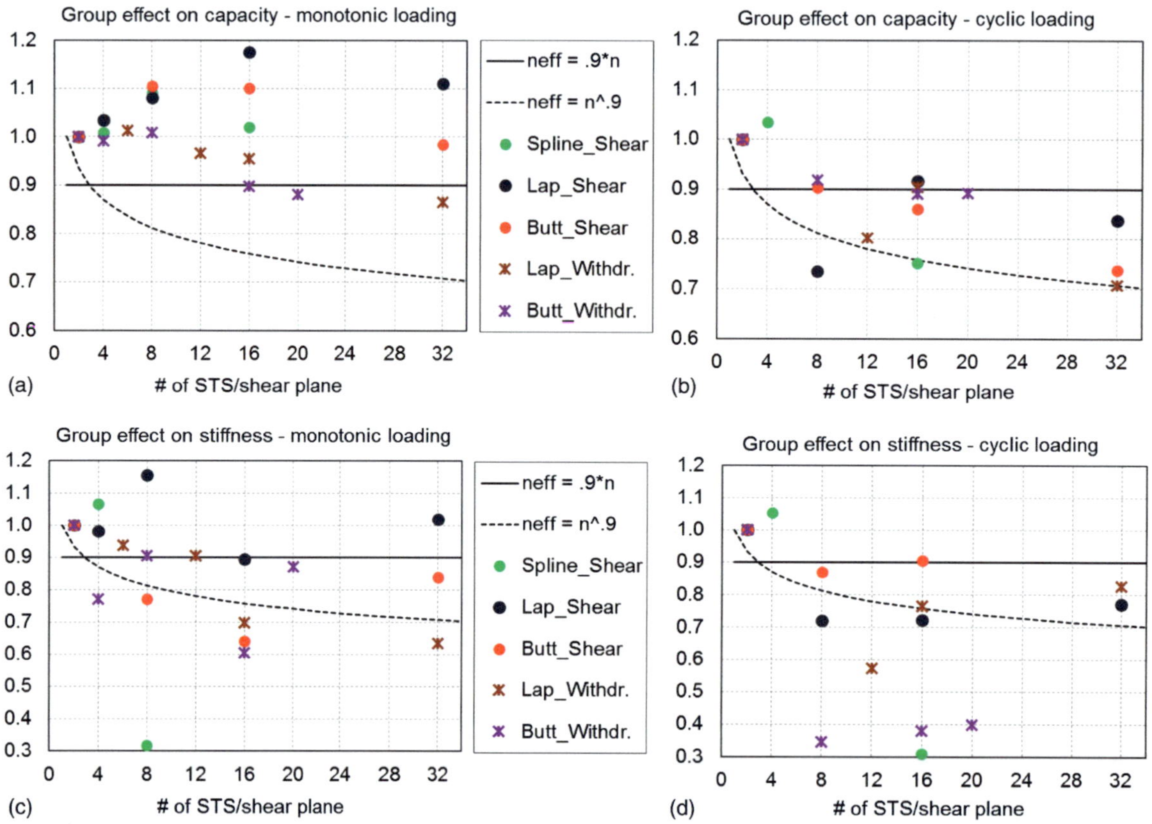

Figure 44. Group effect on (a) monotonic capacity; (b) cyclic loading capacity; (c) monotonic stiffness; and (d) cyclic loading stiffness (Hossain et al., 2019).

Findings related to the group effect on the strength and stiffness of the tested connections appear in Figure 44. Results showed that the expression $n_{eff} = 0.9n$ is sufficient to describe the reduction in capacity for all joints under quasi-static monotonic loading with STSs acting in shear or in withdrawal. This approach is significantly less conservative than the existing Canadian design guidance for lag screws. The strength (capacity) under reversed cyclic loading was on average 9% lower than that from monotonic loading. The study suggested the expression $n_{eff} = n^{0.9}$ to account for the group effect of STS joints under cyclic loading. It is necessary to properly account for this reduction in performance when designing CLT seismic LLRS. Because most manufactures only provide information on the monotonic performance of their screws, it is particularly important to consider these findings for seismic design. The expression $n_{eff} = n^{0.8}$ conservatively describes the effect of the number of screws on the joint stiffness. Due to the high variation within test series and the large differences between them (COV of 27%), the authors of the study suggested further investigation. The stiffness under reversed cyclic loading was on average 9% lower than under monotonic loading for joints with STSs acting in withdrawal, but it did not decrease for STSs in shear. Finally, the group effect for ductility for all joints can be expressed with the equation $n_{eff} = n^{0.9}$ for both monotonic and reversed cyclic loading cases.

Tests conducted at Oregon State University by Sullivan et al. (2018), described in detail in section 5.6, also showed reduction factors for strength in the range of 0.80 to 0.85 for CLT connections with STSs in lap and spline joints. On the other hand, later tests at the same institution by Taylor et al. (2020), which will also be discussed in detail in section 5.6, show no reduction in strength for nailed and screwed connection in CLT. This suggests that more research is necessary. To be on the conservative side, designers should take into consideration the reduction of strength and stiffness given a larger number of screws. This guide would recommend a reduction in the range of 0.8 to 0.9.

5.5 Effect of the Angle of STSs on the Connection Properties

Combining STSs under different installation angles in one connection can help achieve high connection stiffness and ductility. Hanna and Tannert (2021) conducted experimental investigation on glued-laminated timber joints with STSs that included several combinations of screws acting in withdrawal, along with some acting laterally in shear. The connections used Douglas fir glulam 24f-E with properties as per CSA O86 (CSA, 2019). The average moisture content, determined using a handheld resistance moisture meter, was 8.4%, with a COV of 12%. The average apparent density of the glulam was 556 kg/m3. The tests used Heco-Topix®-CC STSs with nominal and effective core diameters of d = 8.5 mm and d_{eff} = 8.0 mm, respectively, and with lengths of 150, 215, and 300 mm. These STSs consisted of two threads with different pitches with a central shank, which helps bring the members closer. The screws' European technical approval ETA-12/0132 (DIBt, 2012) specified the yield moment as M_y = 20.0 Nm.

The study authors tested thirteen series of connection specimens, labelled S1–S13, with different screw configurations (Figure 45). Five replicates were tested for each series, for a total of 65 tests. Each specimen was 550 mm high, 146 mm wide, and 80 mm deep, with a single shear plane as a joint between the two glulam members. Each joint had four screws installed, with various angles between screw axis and shear plane. The tests used different screw lengths based on the joint configuration. All specimens used a spacing parallel to the grain of 50 mm. To avoid interference between inclined screws, there were two rows of screws with a spacing across the grain of 20 mm. The specimens all met or exceeded the minimum required edge and side distances, as per the product approval (DIBt, 2012), in all configurations. The study authors rotated the test specimens for 12°, similar to the procedure suggested in CEN EN-408 (CEN, 2012), to align the resultant force between loading and support. The application of loads followed the guidelines in CEN's EN-26891 (1991), with displacement-controlled loading rates of 3 and 10 mm/min. for the test series exhibiting brittle and ductile behaviours, respectively. This involved loading specimens to 40% of the estimated capacity, unloading them to 10% of estimated capacity, then finally loading them to failure, defined as the point at which the load dropped to 80% of the maximum. The authors analyzed the impact of the screw installation angle on connection performance in terms of the yield load F_Y, maximum load F_{max}, displacement at yield load d_Y, displacement at maximum load d_{Fmax}, stiffness k, and ductility μ. While parameters such as F_{max} were determined directly from the load-displacement curves, k was computed for the range between 10% and 40% of F_{max}, as per CEN EN-26891 (1991). To identify F_Y and d_Y, the 5%-diameter offset method was used (Muñoz et al., 2008), which requires a straight line with a slope equal to the initial stiffness at an offset on the relative displacement axis equal to 5% of the diameter of the fastener.

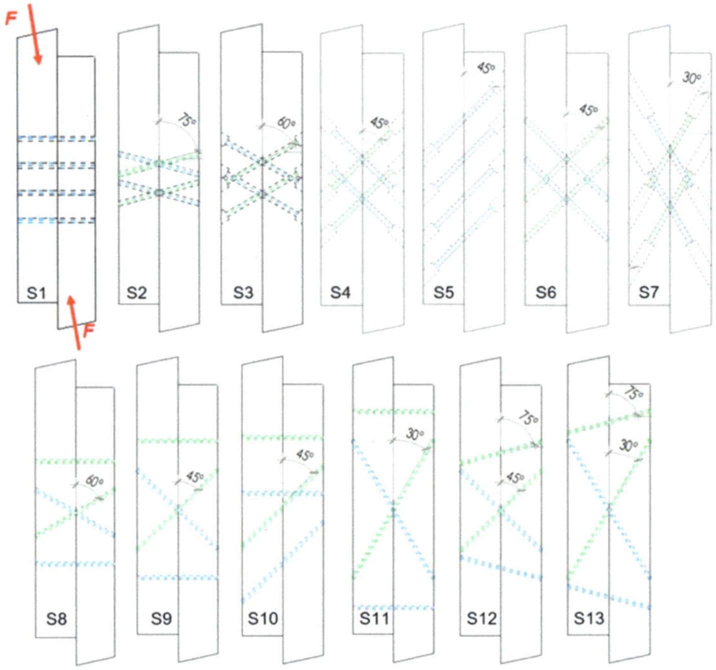

Figure 45. Specimen configuration for all test series (Hanna & Tannert, 2021).

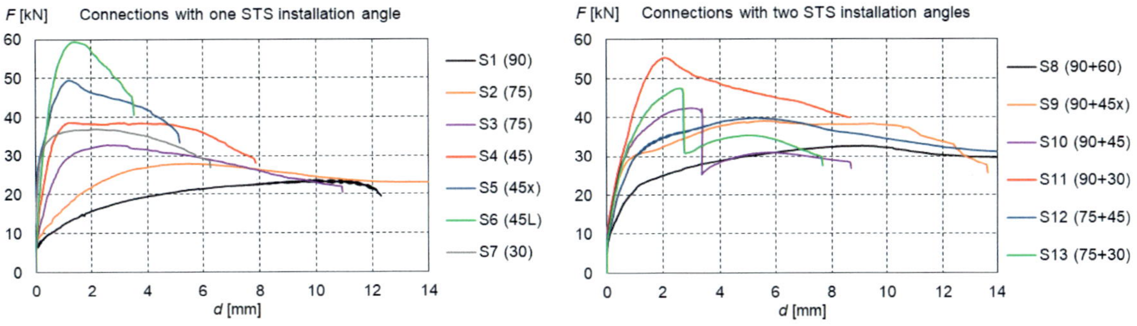

Figure 46. Typical load-deformation curves from test series S1 to S7 (left) and test series S8 to S13 (right) (Hanna & Tannert, 2021).

Load-deformation curves from the testing appear in Figure 46, while the numerical values for the main parameters and percentages of variability are in the table in Figure 47. Hanna and Tannert showed that the load-carrying capacity increased with a decrease in the installation angle, from 25 kN for S1 with screws at 90°, to 40 kN for S4 with screws at 45°, then to 36 kN for S7 with screws installed at 30° (Figure 46 left). When comparing the series with 45-degree screws, there was an approximately proportionate load-carrying capacity increase with increased embedment length, from S4 (40 kN for 70 mm) to S6 (59 kN for 100 mm), and a 25% increase when installing all screws so they acted in shear-tension (S5 with 50 kN). The authors observed the same trends when combining screws at different angles (S8–S13). Variability within tests series was 8% on average, with a maximum of 14% for S8 and a minimum of 3% for S11.

Differences in displacement capacity between test series were much more pronounced, with d_{Fmax} decreasing significantly for smaller installation angles: from 9.0 mm for S1 to 1.2 mm for S4 (Figure 46). Only series S7, with screws at 30°, did not follow this trend, with d_{Fmax} = 3.0 mm. When comparing the series with 45° screws, the authors observed only very small differences, with 1.5 and 1.4 mm for S5 and S6, respectively. Series which combined screws at different angles (S8–S13) reached their displacement capacities between 2.9 mm and 5.0 mm. Only S8, which combined screws at 90° and 60°, showed different behaviour, with d_{Fmax} = 12.5 mm. Variability within the test series was 26% on average, with a maximum of 68% for S8 and a minimum of 12% for S4.

	F_{max}		d_{Fmax}		F_y		k		μ		R	F_{max}/R
	[kN]		[mm]		[kN]		[kN/mm]		[-]		[kN]	[-]
S1	25,1	9 %	9,0	18 %	13,4	10 %	5,2	25 %	7,6	18 %	25,2	1,0
S2	28,5	7 %	4,8	23 %	18,5	5 %	8,1	25 %	3,7	22 %	24,1	1,2
S3	32,8	5 %	3,2	22 %	27,2	6 %	14,7	34 %	4,0	23 %	29,6	1,1
S4	39,8	6 %	1,2	14 %	37,1	9 %	28,4	17 %	1,6	21 %	33,6	1,2
S5	49,5	7 %	1,4	26 %	43,4	9 %	47,5	10 %	2,2	28 %	33,6	1,5
S6	58,7	8 %	1,5	15 %	55,1	10 %	23,5	18 %	1,5	5 %	44,3	1,3
S7	35,7	5 %	3,0	41 %	31,8	8 %	64,9	48 %	3,3	41 %	29,1	1,2
S8	35,3	15 %	8,5	14 %	24,4	8 %	10,1	31 %	6,5	25 %	27,4	1,3
S9	40,4	6 %	3,9	33 %	31,0	6 %	17,7	40 %	4,1	27 %	34,8	1,2
S10	43,5	3 %	2,9	15 %	35,0	9 %	12,9	35 %	2,7	26 %	34,8	1,3
S11	52,3	14 %	3,3	47 %	43,9	10 %	16,9	10 %	3,3	46 %	39,8	1,3
S12	41,0	11 %	5,0	19 %	31,3	13 %	13,3	27 %	5,5	21 %	34,2	1,2
S13	48,6	10 %	4,9	21 %	38,5	5 %	17,1	37 %	5,3	20 %	39,3	1,2

Figure 47. Summary of the test results, with the percentages of variability (Hanna & Tannert, 2021). Note: Commas in the numbers in the table are decimal commas.

Connection stiffness k increased with a decrease in the installation angle, from k = 4.5 kN/mm for S1 to k = 67 kN/mm for S7 (screws at 30°). When comparing series with 45°degree screws, results showed a large increase in stiffness when installing all screws to act in shear-tension (S5 with 41 kN/mm). The series combining screws at different angles (S8–S13) exhibited stiffness between 10 and 30 kN/mm for S8 and S13, respectively. There was no clear correlation between the stiffness results for screws installed at a single angle vs. those installed at two different angles. Variability within the test series for stiffness was large: 49% on average, with a maximum of 98% for S12 and a minimum of 15% for S4.

Ductility followed a very similar trend as the deformation capacities, with μ decreasing significantly with a decrease in installation angle, from μ>6 for S1 to μ = 1.5 for S4. Once again, series S7, with STSs at 30°, was an exception to this trend with μ = 4.4. When comparing the different series with 45° screws, despite some differences, all exhibited low ductility (μ<2.0). Test series combining screws at different angles (S9–S13) achieved moderately ductile behaviour, while S8, combining screws at 90° and 60°, was ductile with μ > 2.0.

5.6 Half-Lap and Spline Connections Using STS in CLT Conducted at Oregon State University

Sullivan et al. (2018) investigated the properties of CLT diaphragm connections and compared them to the design values calculated from formulas in timber design standards such as the NDS (American Wood Council [AWC], 2015) and the Eurocode (CEN, 2004a). They tested six different specimen configurations, each specimen consisting of three separate CLT panels (each 2.44 × 0.61 m) connected side-by-side along the long edge. CLT panels were three-ply, V1 grade Douglas fir-Larch (DF-L) panels, with DF-L No.3 in the core layer and DF-L No.2 in the outer layers. The authors tested half-lap and surface spline connections with 8 and 10 mm STSs spaced at 152 mm (6 in.) and 305 mm (12 in.). These tests used the SWG ASSY 3.0 Eco model and VG CSK model screws.

Three of the four surface spline connections used SWG ASSY Eco (partially threaded and counter-sunk) screws, with SWG ASSY VG CSK fully threaded screws for one surface spline connection and all the half-lap connections. Edge and end spacing for the DF-L CLT were 7d and 22.5d, respectively, as per ICC-ES Report ESR-3179 for SWG ASSY 3.0 Wood Screws. Figure 48 shows the test matrix; "SS" means a surface spline connection while "HL" means a half-lap connection.

Construction #	Test #	Load type	Connection type	Thread type	Spacing	Screw dia.	Δ_{ref} (mm)
1	1	Mono.	SS	Partial	152 mm	8 mm	--
1	2	Mono.	SS	Partial	152 mm	8 mm	--
1	3	Cyc.	SS	Partial	152 mm	8 mm	24
1	4	Cyc.	SS	Partial	152 mm	8 mm	81
2	5	Mono.	SS	Partial	152 mm	10 mm	--
2	6	Mono.	SS	Partial	152 mm	10 mm	--
2	7	Cyc.	SS	Partial	152 mm	10 mm	41
2	8	Cyc.	SS	Partial	152 mm	10 mm	41
3	9	Mono.	SS	Partial	305 mm	10 mm	--
3	10	Mono.	SS	Partial	305 mm	10 mm	--
3	11	Cyc.	SS	Partial	305 mm	10 mm	41
3	12	Cyc.	SS	Partial	305 mm	10 mm	41
4	13	Mono.	SS	Full	152 mm	8 mm	--
4	14	Mono.	SS	Full	152 mm	8 mm	--
5	15	Mono.	HL	Full	152 mm	10 mm	--
5	16	Mono.	HL	Full	152 mm	10 mm	--
5	17	Cyc.	HL	Full	152 mm	10 mm	15
5	18	Cyc.	HL	Full	152 mm	10 mm	30
6	19	Mono.	HL	Full	305 mm	10 mm	--
6	20	Mono.	HL	Full	305 mm	10 mm	--
6	21	Cyc.	HL	Full	305 mm	10 mm	15
6	22	Cyc.	HL	Full	305 mm	10 mm	15

Figure 48. Test matrix for the various diaphragm connection configurations (Sullivan, 2017); SS = Surface Spline; HL = Half-Lap.

Two replicates per connection configuration under static and cyclic loading were tested based on the requirements given in ASTM E455 (ASTM International, 2011) and ASTM E2126 (ASTM International, 2002). Figure 49 shows perspective views of the surface spline (right) and half-lap (left) specimen types. Figure 50 contains detailed information on each construction type. Figure 49 shows the unique WSSW install pattern (**w**ithdrawal/**s**hear/**s**hear/**w**ithdrawal) used in the specimens. In this case, screws placed near either end of the connection were at 45° in the direction of the loading, thus contributing to the shear capacity of the connection in either screw withdrawal or compression, depending on the direction of the shear force. The screws in the middle of the joint acted in pure shear, being installed vertically. A 25.4 mm (1 in.)–thick plywood provided the surface splines. Moisture content proved to be between 5% and 10% for the plywood and CLT in all specimens within 30 minutes of testing.

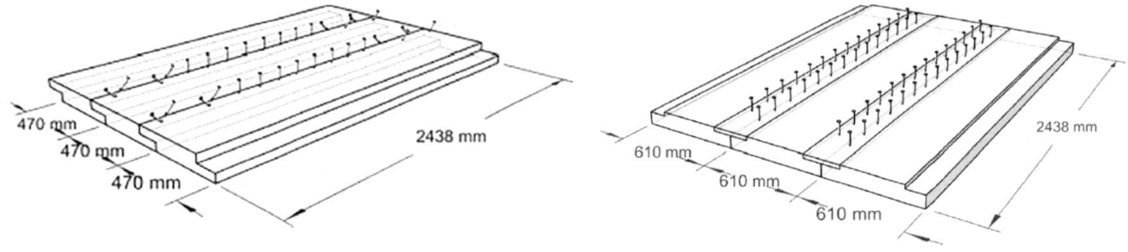

Figure 49. Perspective view of the half-lap (left) and spline (right) connection specimens (Sullivan et al., 2018).

Type	Construction #	Test#	Load Type	# of fasteners	Screw diameter	Test peak load	Peak load per fastener	Peak unit shear load	EEEP yield load	Displacement at peak load	Displacement at EEEP yield	Ductility	Stiffness
					D_n	V_{max}			V_Y	Δ_{max}	Δ_y	μ	k
					(mm)	(kN)	(kN)	(kN/m)	(kN)	(mm)	(mm)		(kN/mm)
Column ID	1	2	3	4	5	6	7	8	9	10	11	12	13
Spline	1	1	M	28	8	181	6.5	74	161	51	13.5	3.8	10.9
↓	1	2	M	28	8	179	6.4	73	161	47	12.3	3.8	11.1
	1	3	C	28	8	135	4.8	55	117	24	7.3	3.2	15.3
	1	4	C	28	8	164	5.9	67	151	23	9.6	2.4	15.7
	2	5	M	26	10	219	8.4	90	196	77	26.4	2.9	6.8
	2	6	M	26	10	229	8.8	94	203	50	18.7	2.7	10.2
	2	7	C	26	10	209	8.0	86	176	28	9.2	3.0	18.5
	2	8	C	26	10	240	9.2	98	206	28	8.8	3.2	22.7
	3	9	M	14	10	141	10.1	58	126	54	20.9	2.6	5.0
	3	10	M	14	10	144	10.3	59	127	73	19.9	3.6	6.1
	3	11	C	14	10	118	8.4	48	99	28	7.3	3.8	13.3
	3	12	C	14	10	116	8.3	48	103	28	8.0	3.5	12.7
	4	13	M	28	8	246	8.8	101	205	31	17.0	1.8	11.9
	4	14	M	28	8	225	8.0	92	190	50	17.3	2.9	9.2
Half-lap	5	15	M	8	10	279	34.8	114	263	16	9.2	1.7	29.7
↓	5	16	M	8	10	245	30.6	100	237	10	7.6	1.4	30.5
	5	17	C	8	10	272	34.0	111	248	9.4	5.6	1.7	44.0
	5	18	C	8	10	281	35.2	115	264	20	6.3	3.2	42.0
	6	19	M	4	10	138	34.6	57	128	10	7.9	1.2	17.3
	6	20	M	4	10	173	43.2	71	155	7.1	3.5	2.1	43.1
	6	21	C	4	10	136	33.9	56	125	5.7	3.5	1.6	35.7
	6	22	C	4	10	138	34.5	57	130	5.9	4.1	1.4	31.6

Figure 50. Results summary (Sullivan, 2017).

The results from the experimental tests summarized in Figure 50 are valuable to designers as mass timber diaphragms often use these connections. The calculation of the parameters used the EEEP methodology. Peak loads for surface spline connections with 25 mm–thick splines averaged 5.9 kN per screw for 8 mm diameter screws and 8.9 kN per fastener for 10 mm screws. Half-lap capacities averaged 35.1 kN per fastener for 10 mm screws. Stiffness values for surface splines varied from 0.36 to 0.87 kN/mm per fastener. Stiffness values for half-laps varied from 4.3 to 5.5 kN/mm per fastener. The stiffness values per fastener are not provided in Figure 50 but are available in Sullivan (2017) and Sullivan et al. (2018). While the NDS does not provide formulas to estimate stiffness, equations in EN-1995 estimate 8 mm diameter screws in Douglas fir lumber in pure shear to have a stiffness of 3.9 kN/mm.

The load-deflection curves (not shown here) of spline specimens exhibited high ductility, as the screw heads bend and pull through the plywood spline. The WSSW installation pattern for the half-lap connections exhibited high ductility due to the vertical screws in pure shear. Strength per screw increased alongside the cross-sectional area of the screw and did so in an approximately proportional fashion. Also, the 8 mm screws exhibited higher ductility than the 10 mm screws. This information can be useful if one considers it in the development of a new diaphragm force reduction factor for CLT diaphragms, as steel fasteners in wood are the main source of energy dissipation during earthquakes.

In the monotonic tests, group factors that depend on the increase in the number of fasteners ranged from 0.8 to 0.85, calculated using the Eurocode, the New Zealand Timber code, and the NDS. Monotonic surface spline specimens initially yielded in Mode III_s per NDS, in combination with a screw head pull-through the plywood thereafter. On the other hand, the initial yield of the monotonic half-lap specimens was related to the withdrawal of the angled screws, subsequentially followed by Mode III_m failure for the vertical screws in shear. Small cycle fatigue in the screws was the main mode of failure for the spline specimens under cyclic loads. The screws in the half-lap connections remained intact during the cyclic loading, and the failure thereafter was characterized by a withdrawal of the angled screws, along with wood crushing around the vertical screws (Mode III_m).

Taylor et al. (2020) investigated the in-plane shear behaviour of CLT panel-to-panel surface spline connections using nails and screws at different fastener spacings. The connection specimens had two shear planes, as shown in Figure 51. This figure presents a plan and a cross-sectional view of the tested connections, while the test matrix appears in the table in Figure 52.

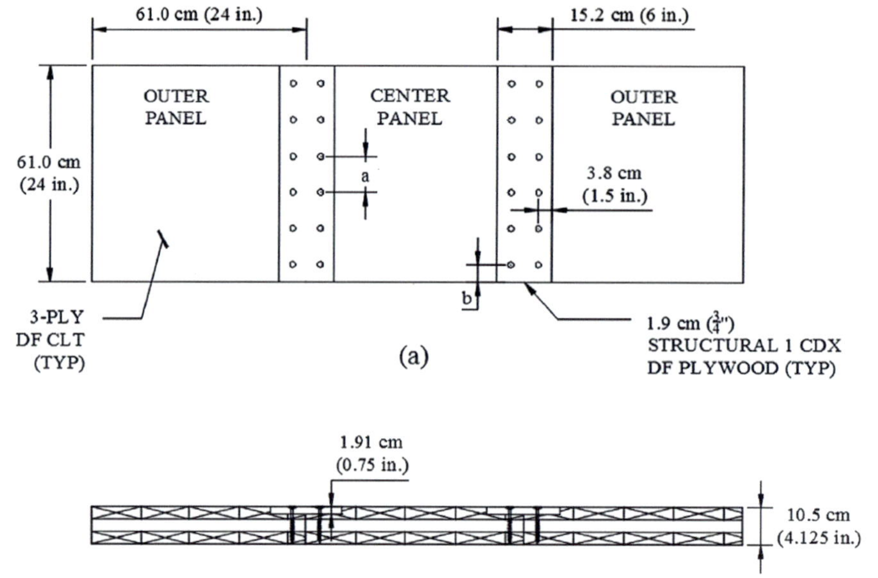

Figure 51. Plan and cross-sectional view of the tested connections (Taylor et al., 2020)

Series ID	Number of Trials	Fastener Type	Fastener Spacing, a [cm *(in.)*]	End Distance, b [cm *(in.)*]
Nail@3"	5	16d Common Nail	7.62 *(3)*	3.81 *(1-½)*
Screw@8"	5	SDWS22400DB-R50*	20.3 *(8)*	10.2 *(4)*
Screw@6"	5	SDWS22400DB-R50*	15.2 *(6)*	7.62 *(3)*
Screw@4"	5	SDWS22400DB-R50*	10.2 *(4)*	5.08 *(2)*

Figure 52. Test matrix for spline and lap connection tests (Taylor, 2019) * More information on screws is available at https://www.strongtie.com.

Monotonic tests of each spline configuration produced similar damage mechanisms. The initial damage was bending of the fasteners (Figure 53 right), followed by pull-through as the head of the fasteners plunged into the top ply of the plywood. The fasteners then began to yield in mode III_s, causing a reduction in the specimen stiffness at points 1N and 1S. The III_s damage mode is characterized by bearing in the plywood and the bending of the fastener, with a single-fastener plastic hinge forming at the first layer of CLT. This mode remained unchanged until the specimen reached its peak force at points 2N and 2S in Figure 53 left. As the panel experienced post-peak force deformations (point 2N) and the strength of the system started to degrade, the nail heads simply pulled through the plywood, causing minimal bearing damage (Figure 53 right). At an increased displacement level (point 3N), the nail heads began to completely pull through the plywood and nail bearing damage in the CLT became evident (Figure 53 right). During post-peak force deformations (point 3S), the screw specimens displayed larger degrees of plywood damage caused by the fastener, which increased as the screw spacing decreased. This arises from the increasing force capacity of the system thanks to the increased number of screws. Removing the plywood spline at the end of testing revealed screw bearing damage in the CLT as well.

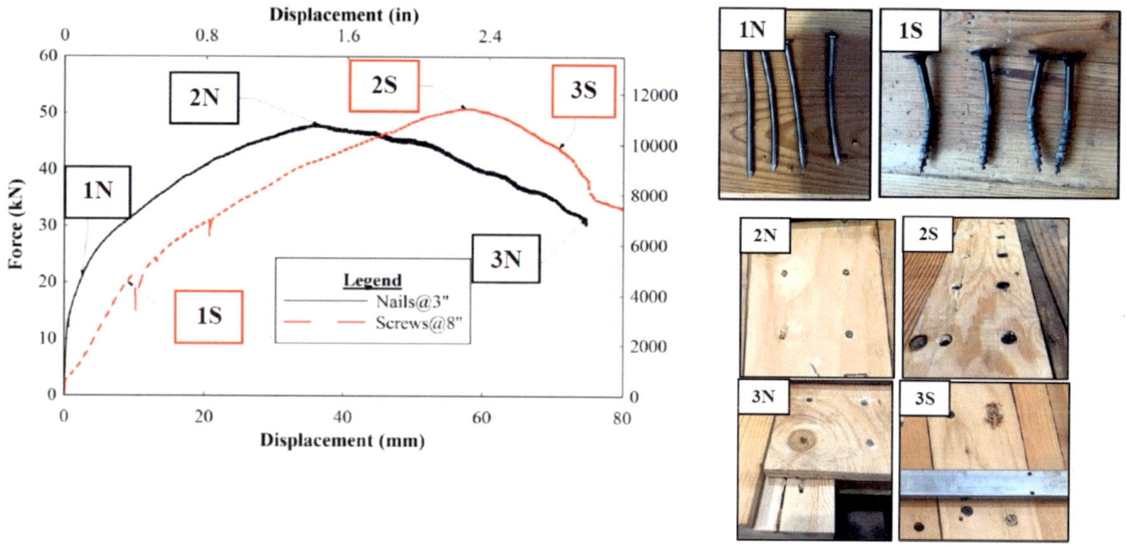

Figure 53. Plan and cross-sectional views of the tested connections (Taylor et al., 2020).

The cyclic tests showed similar damage mechanisms to those in the monotonic tests. Elastic bending of the fasteners (point 1) was followed by their mode III$_s$ yielding (point 2). The controlling failure point (point 3) of the screws was always due to low-cyclic fatigue causing sudden screw fractures right above or below the thread line (Figure 53 right). The screws caused varying degrees of plywood damage during post-peak displacement cycles, followed by a head pull-through. Point 4 in Figure 54 shows the residual strength after the first screw fatigue failure (Point 3, Figure 54); not all of the screws fatigue failed during the same displacement cycle. For the nailed specimens, a gradual combination of nail pullout and/or low-cyclic fatigue failure occurred during post-peak force deformations (Point 3, Figure 54). There were also varying degrees of nail bearing damage in the plywood. The failure mechanism for the nails appeared less sudden than for the screws.

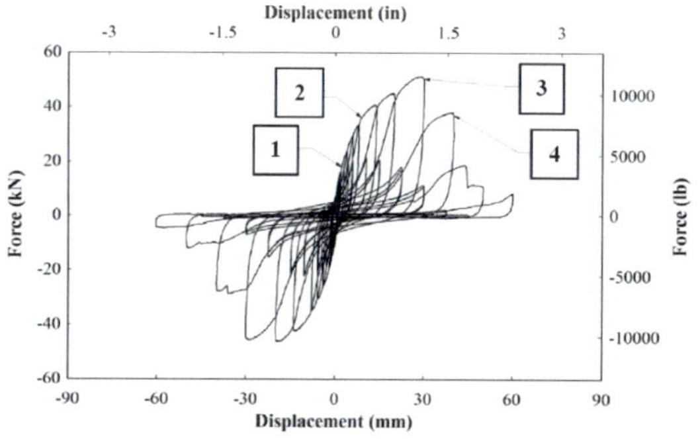

Series ID	K [kN/mm] (lb./in.)	F_y [kN] (lb.)	F_{max} [kN] (lb.)	F_{ult} [kN] (lb.)	Nominal Shear Capacity [kN] (lb.)	Ω
Nail@3″	0.177 (1010)	0.530 (1 1 9)	1.31 (2 9 5)	1.13 (2 5 5)	0.93 (2 0 8)	1.42
Screw@8″	0.334 (1900)	1.59 (3 5 8)	3.62 (8 1 5)	3.31 (7 4 3)	1.35 (3 0 3)	2.69
Screw@6″	0.308 (1760)	1.56 (3 5 1)	3.44 (7 7 5)	3.02 (6 7 9)	1.35 (3 0 3)	2.56
Screw@4″	0.294 (1680)	1.57 (3 5 2)	3.53 (7 9 3)	3.09 (6 9 5)	1.35 (3 0 3)	2.62

Figure 54. Typical hysteretic curve obtained during testing of connections with STSs (left) and the connection strength and stiffness parameters per fastener (Taylor et al., 2020).

The elastic stiffness K is the slope of the line between 10% and 40% of the maximum load F_{max}, while the calculation of the yield force F_y and yield deformation D_y used the 5% offset method. The average elastic stiffness K of the connections with screws ranged from 4.01 kN/mm to 7.06 kN/mm and, as expected, increased as the screw spacing decreased. The nail specimens displayed an average stiffness of 5.66 kN/mm, a 14.8% increase and 19.8% decrease, respectively, compared to the screw specimens with 6 and 4 in. spacing. There were no statistically significant differences between stiffness per screw mean values. The average stiffness per nail was 0.177 kN/mm, a decrease of 43.3% from the average stiffness per screw. The screw specimens displayed an average F_y ranging from 19.1 to 37.6 kN and, as expected, increased as the screw spacing decreased. Nailed specimens had an average F_y of 17.0 kN, which was most comparable to the 8 in. screw spacing. There were no statistically significant differences between yield force per screw mean values. Nailed specimens showed a 66.3% decrease in the average yield force per fastener compared to the average F_y of the screw specimens. The average yield displacement D_y ranged from 2.3 mm for nails to 4.3 mm for screws with 4 in. spacing. The average F_{max} for the screw specimens varied from 43.5 to 84.7 kN. The average F_{max} for the nailed specimens was 41.9 kN, comparable to that of connections with 8 in. screw spacing. The maximum load per fastener, F_{max}, for the nail specimens decreased, on average, by 63% compared to the equivalent average F_{max} for the screwed specimens. The average maximum deformation D_{max} showed no significant trend amongst fasteners and varied from 28.4 to 33.8 mm. The table in Figure 54 right summarizes the force-based parameters for each fastener.

The average ductility (μ) ranged from 9.5 to 16.5, with the nailed specimens displaying the largest ductility, which was 58.7% higher than the average ductility of the screwed specimens. The average overstrength (Ω) values for the screw specimens ranged from 2.56 to 2.69. The average overstrength factor for the nail specimens was 1.42, a 46% decrease from the average of the screwed specimens. The increase in ductility is partly attributable to the substantial increase in overstrength for the screws. Compared to the results of similar tests by Sullivan et al. (2018) and Hossain et al. (2018), the average stiffness per screw in Taylor et al. (2020) was 43.3% and 65.3% lower, respectively. The average peak force per screw was also 26.5% and 30.8% lower than the results in those two studies, respectively. Such decreases in strength and stiffness were likely due to the use of STSs with roughly half the root diameter of those used by Sullivan et al. and Hossain et al.

5.7 Half-Lap and Spline Connections in CLT Conducted at Texas A&M University

Jalilifar et al. (2021) conducted a total of 56 tests on three common types of CLT shear connection, namely surface spline (SS), half-lap (HL), and CLT-to-glulam butt joints at Texas A&M University, investigating the response of each type of connection with varying configurations under both monotonic and cyclic loading. Cyclic loading was applied using the Consortium of Universities for Research in Earthquake Engineering (CUREE, 2000) testing protocol. The testing matrix appears in Figure 55. For each specimen configuration (specimen number, SN) in Figure 55, a series of two monotonic and five cyclic loading tests were performed.

SN	Connection type	Fastener	Fastener orientation	Fastener length (in.)	Fastener spacing (in.)
1	Surface spline[a]	16d common nail	90	3.5	3
2	Surface spline[a]	16d common nail	90	3.5	6
3	Half-lap[a]	16d common nail	90	3.5	3
4	Surface spline[a]	MyTiCon ASSY Eco	90	4	6
5	Surface spline[a]	Simpson SDWS22400DB	90	4	6
6	Surface spline[a]	Simpson SDWS22400DB	45	4	6
7	Half-lap[a]	Simpson SDWS22400DB	90	4	6
8	Butt-joint[b]	Simpson SDWS22800DB	90	8	6

Figure 55. Test matrix CLT connections (Jalilifar et al., 2021).

The selection of connection configurations was based on common practices for CLT diaphragms. The HL and SS connections consisted of three-layered CLT panels made of Southern Pine with dimensions 610 × 305 × 105 mm and two shear planes. The adjacent CLT panels in the SS connections used one 28.6 mm–thick plywood and 16d common nails with 76 mm (3 in.) or 152 mm (6 in.) spacing. Some SS connections used MyTiCon ASSY 5.8 x 100 mm STSs, while others used Simpson SDWS22400DB, D = 5.5 x 102 mm (0.22 x 4 in.) or 203 mm (8 in.) screws with 152 mm (6 in.) spacing.

HL connections had a 152 mm (6 in.)–wide overlap to provide appropriate edge distance for the fasteners. Two series of HL tests were conducted with one row of 16d nails with 76 mm (3 in.) spacing or Simpson SDWS22400DB screws with 152 mm (6 in.) spacing. SS connection specimens SN1 had 10 nails (5 per shear plane), while SS connections SN2 had 4 nails (2 per shear plane). HL connection SN3 had 5 nails per shear plane, SS connections SN4 had 4 MyTiCon screws (2 per shear plane), and SS connections SN5 had 4 Simpson screws (2 per shear plane). Finally, the SS connections SN6 had 4 Simpson screws (2 per shear plane), while HL connection SN7 had 2 Simpson screws per shear plane.

SN	Test	Peak load (kips)	Δ_{ref} (in.)	Peak load per fastener (kips)
1	M1	5.07	1.79	0.51
	M2	5.00	1.30	0.50
2	M1	1.86	1.25	0.47
	M2	2.01	1.40	0.50
3	M1	4.74	0.46	0.47
	M2	4.79	0.43	0.48
4	M1	7.55	1.83	1.89
	M2	7.81	1.53	1.95
5	M1	9.31	1.90	2.33
	M2	9.64	2.08	2.41
6	M1	4.46	1.87	1.12
	M2	3.06	1.71	0.77
7	M1	7.01	0.76	1.75
	M2	6.46	0.75	1.62
8	M1	7.50	0.83	1.88
	M2	8.86	0.81	2.22

Figure 56. Peak load, reference displacement, and peak load per fastener from the monotonic tests (Jalilifar et al., 2021).

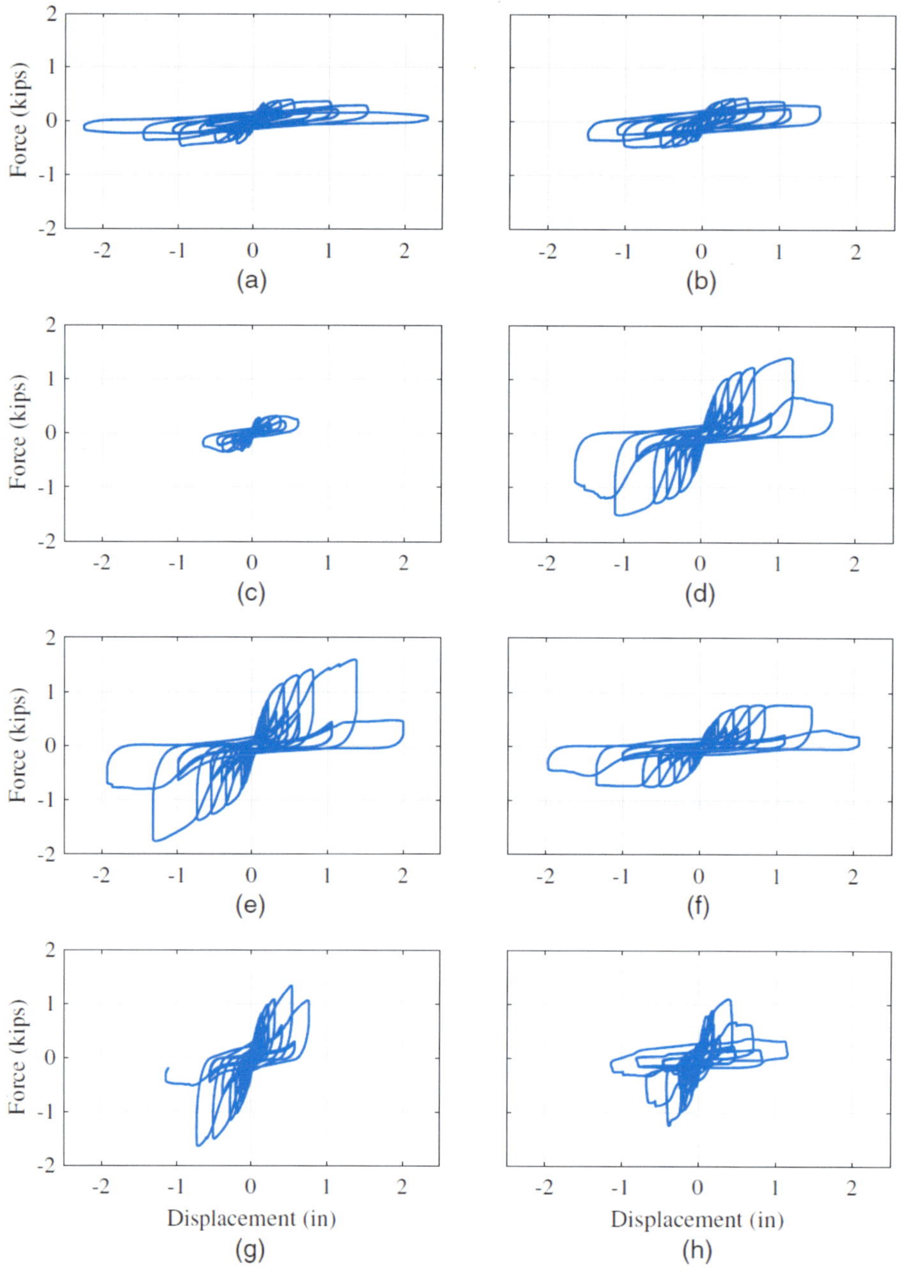

Figure 57. Average force-displacement curve per fastener for the tested CLT shear connections: (a) SN 1; (b) SN 2; (c) SN 3; (d) SN 4; € SN 5; (f) SN 6; (g) SN 7; and (h) SN 8 (1 kip = 4.45 kN, 1 in. = 25.4 mm) (Jalilifar et al., 2021).

The results showed that under monotonic loading, screwed connections had a maximum load-carrying capacity per fastener up to five times higher than the nailed connections, a ratio that reached 1.75 times more per fastener for the cyclic tests (Figure 56). The higher capacity of screwed connections may be attributable to the higher strength of the STSs and the tension developed in the screw from its heads. Although nailed connections seem to have higher ductility, screwed connections had maximum displacement capacities up to 1.5 times higher than the nailed connections (Figure 57). The failure in the SS connections was in general more ductile than in HL connections with similar fastener types and spacing. The cyclic tests revealed that the SS connections can tolerate higher loads than the HL connections. The SS-to-HL connection strength per fastener ratio was 1.19 and 1.43 for the screwed and nailed connections, respectively. In addition, the SS connections tolerated 2.65 and 3.47 times the maximum displacement of the HL connections for screwed and nailed connections, respectively. The study also found that for fasteners in SS connections driven in at 45°, the peak load decreased by as much as 50% compared to those at a 90° angle. The screw head diameter was very effective in dissipating energy capacity and a small change in the head diameter could increase the dissipated energy up to 37%. Additionally, changing the connection type from HL to SS could increase the energy dissipation up to 250% with the same fastener and spacing. Finally, it was shown that the NDS equations for predicting lateral design values, Z, predicted the design values for screwed SS and HL connections to be from 3 to 5.5 times lower than the ultimate values obtained from testing, depending on the loading type. The NDS equations, however, estimated the failure modes accurately in most cases.

6 EXPERIMENTAL AND ANALYTICAL PERFORMANCE OF MASS TIMBER DIAPHRAGMS

This section provides up-to-date information on experimental and analytical research on mass timber diaphragms. This data will provide designers with important knowledge about the strength, stiffness, and ductility of certain tested and analyzed configurations of mass timber diaphragms.

6.1 Tests on Diaphragms at FPInnovations

To better understand the behaviour of mass timber diaphragms, FPInnovations has conducted tests on seven different mass timber diaphragms with different configurations. All assemblies were 7315 mm in length and 2438 mm in depth. Each configuration consisted of a glulam frame made of glulam members, 300 x 89 mm in cross section, with decking attached to them. The glulam frame consisted of perimeter frame members and one to three middle frame members, depending on the configuration. The first five diaphragm assemblies consisted of glulam decking panels, whereas the remaining two used CLT panels. The assemblies were tested vertically, simply supported, with the load evenly distributed over four points of application. These tests measured displacements at mid-span of the diaphragm on the face in tension for both the framing and the decking using laser displacement sensors. Load and displacement of the actuator were also recorded. Figure 58 shows the test setup.

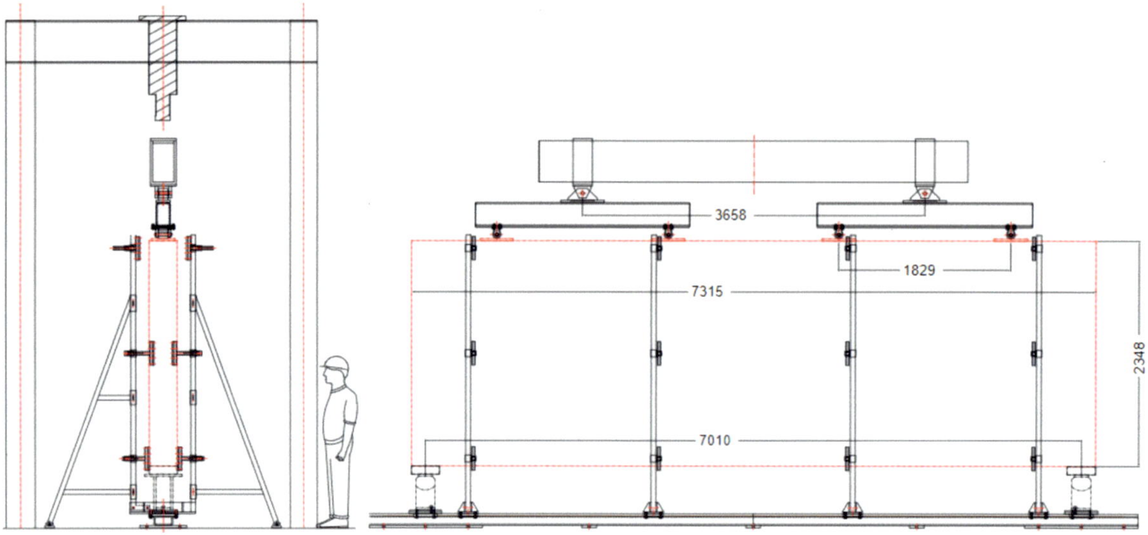

Figure 58. Test setup for determining the in-plane load-deformation behaviour of mass timber diaphragms.

6.1.1 Configuration 1 (GL-89)

Diaphragm configuration 1 used 89 mm–thick glulam beams placed on the flat (glulam panels) as decking. The decking elements were continuous throughout the entire (7315 mm) length of the diaphragm, spanning two middle frame members. The panel-to-panel connection was of a half-lap type manufactured using power-driven nails (83 mm in length and 3.9 mm in diameter). The nails were spaced at 102 mm o.c. and inserted at an angle of 30°. The panel-to-frame connection used ϕ8 mm x 160/80 (diameter x length/thread's length) STSs. There were two screws per panel at every overlapping panel/frame, placed at 152 mm o.c., except on the chord, where screw spacing was 203 mm o.c., as shown in Figure 59.

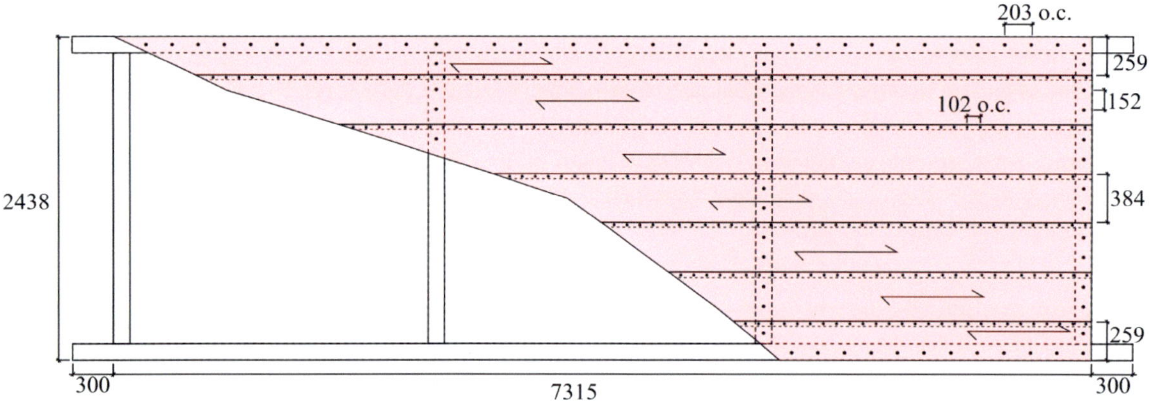

Figure 59. Diaphragm configuration 1: 89 mm–thick glulam panels continuous along the diaphragm's length.

6.1.2 Configuration 2 (GL-89S)

Diaphragm configuration 2 was the same as diaphragm configuration 1, except that the decking panels were arranged in a staggered pattern, as shown in Figure 60. The connections in this diaphragm configuration were the same as those in diaphragm configuration 1.

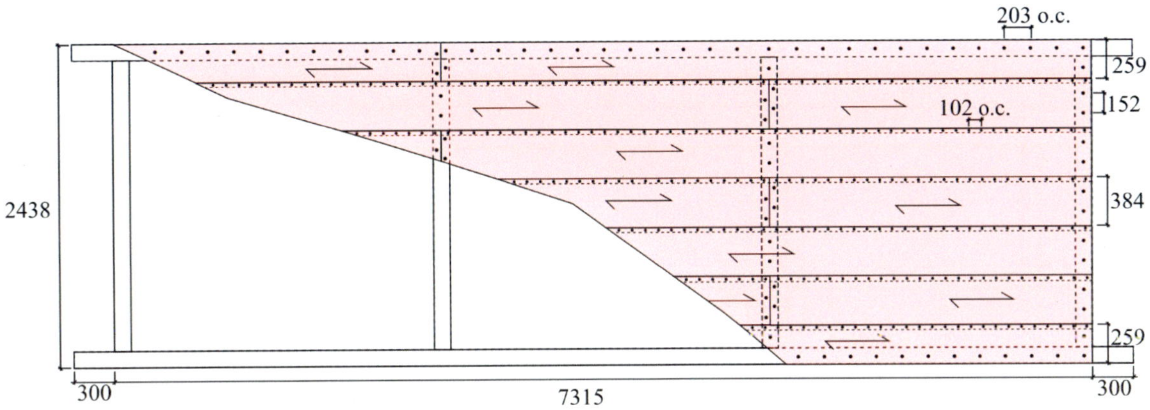

Figure 60. Diaphragm configuration 2: 89 mm–thick glulam panels arranged in a staggered pattern.

6.1.3 Configuration 3 (GL-44)

Diaphragm configuration 3 used 44 mm–thick glulam panels as decking. The decking elements were continuous throughout the 7315 mm length of the diaphragm, spanning three middle frame members. The panel-to-panel connection was half-lap manufactured using power-driven nails (2.5 mm in diameter and 44 mm in length). The nails were spaced at 102 mm o.c. and inserted at an angle of 30° from the horizontal axis. The panel-to-frame connection was manufactured using ϕ6 mm x 100/54 (diameter x length/thread's length) STSs. There were two screws per glulam panel for every overlapping panel/frame, all at 152 mm o.c. except on the chord, where screw spacing was 152 mm o.c., as shown in Figure 61.

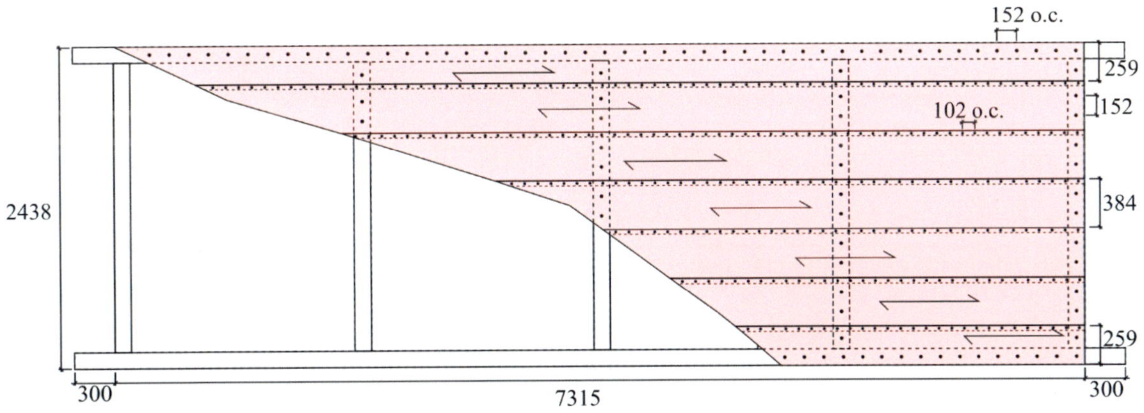

Figure 61. Diaphragm configuration 3: 44 mm–thick glulam panels continuous throughout the diaphragm's length.

6.1.4 Configuration 4 (GL-44S)

Diaphragm configuration 4 was the same as diaphragm configuration 3, except that the decking panels were arranged in a staggered pattern, as shown in Figure 62. The connections for this diaphragm configuration were the same as those in diaphragm configuration 3.

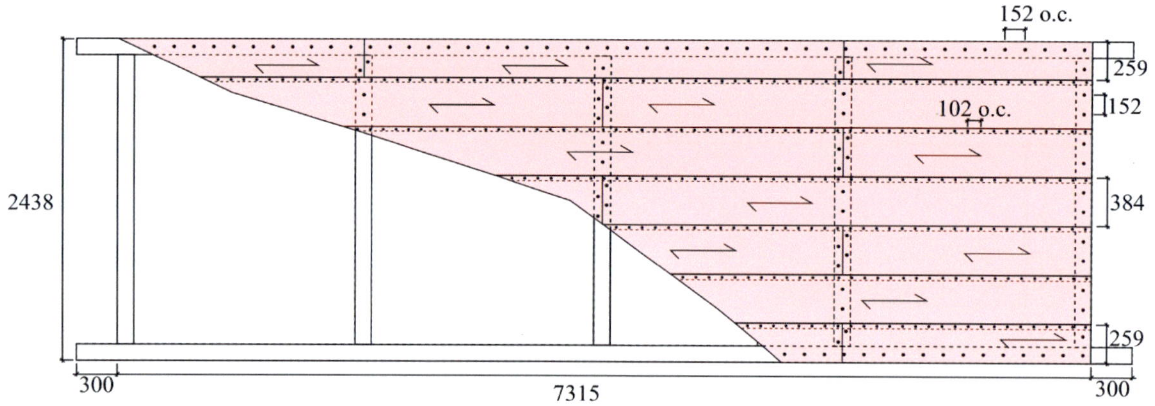

Figure 62. Diaphragm configuration 4: 44 mm–thick glulam panels arranged in a staggered pattern.

6.1.5 Configuration 5 (GL-44P)

Diaphragm configuration 5 used 44 mm–thick glulam panels assembled perpendicularly to the length of the diaphragm, as shown in Figure 63. The connections for this diaphragm configuration were the same than those in diaphragm configurations 3 and 4. This configuration had one middle glulam frame member along the entire diaphragm length.

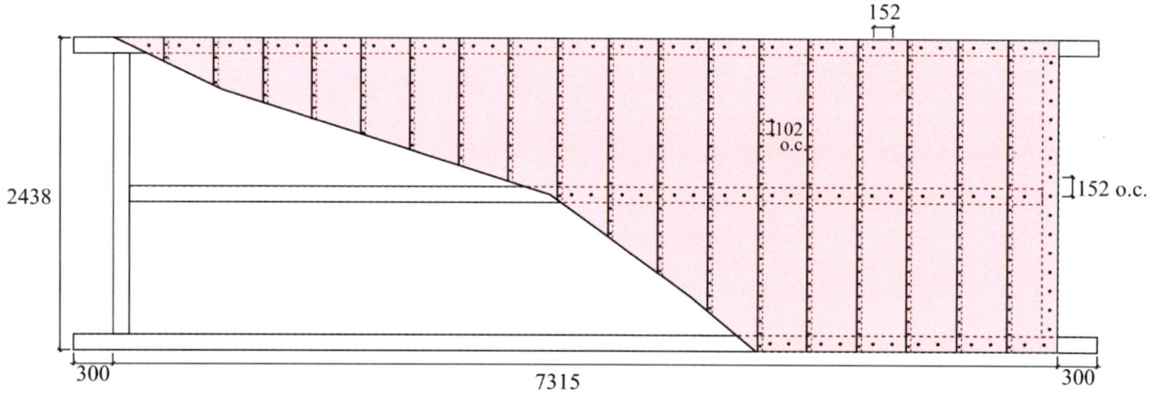

Figure 63. Diaphragm configuration 5: 44 mm–thick glulam panels placed perpendicular to the diaphragm length.

6.1.6 Configuration 6 (CLT-105)

Diaphragm configuration 6 involved a continuous 3-ply (105 mm thick) CLT panel with its major strength direction along the length of the diaphragm, spanning two middle frame members. Since the CLT panel was continuous through the diaphragm's surface, there were no panel-to-panel connections. The panel-to-frame connections used ϕ8 mm x 180 PT (80 thread length) STSs. The STSs were spaced 203 mm o.c. on the perimeter and 406 mm o.c. on the decking surface, as shown in Figure 64.

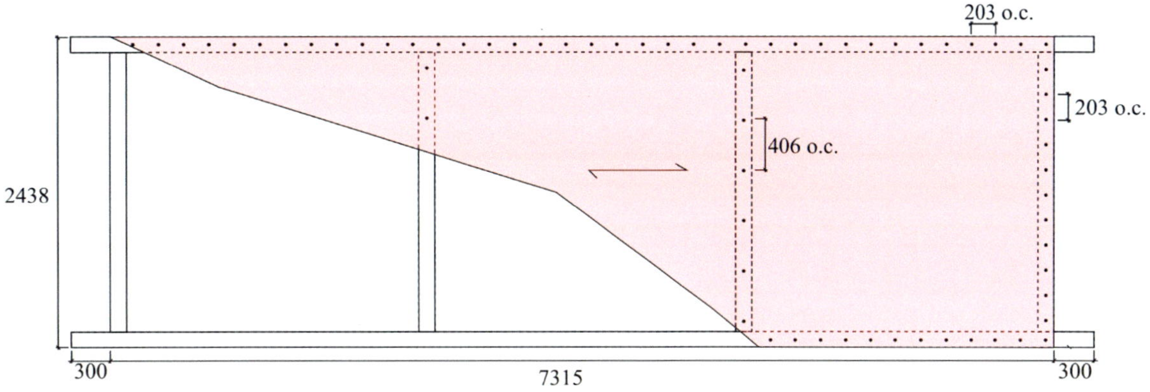

Figure 64. Diaphragm configuration 6: 105 mm–thick 3-ply CLT panel.

6.1.7 *Configuration 7 (CLT-105SP)*

Diaphragm configuration 7 was the same as diaphragm configuration 6, except that it spliced the CLT panels over the two middle frame members, as shown in Figure 65. Panel-to-panel connections utilized a half-lap with the same connection pattern as in diaphragm configuration 6, except that the decking surface used ϕ6 mm x 180 PT (64 thread length) STSs.

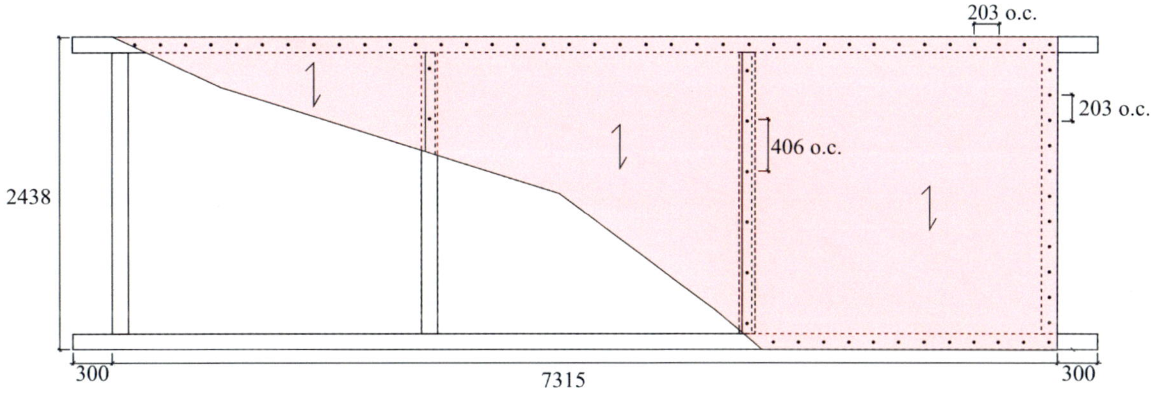

Figure 65. Diaphragm configuration 7: 105 mm–thick 3-ply CLT panels fastened side-by-side.

6.1.8 *Results and Discussion*

All tested diaphragms exhibited nonlinear behaviour. To compare the stiffness and deflection properties of the diaphragms, the stiffness at a deflection of 6.8 mm (L/1080) was used, along with the stiffness estimated at 40% of the maximum load, preloading each diaphragm to around 10 to 20 kN. The deflection at this preload was estimated to plot and analyse the load-deflection curve. The mid-span deflection was measured on the frame and on the decking on the tension side and the average values between the two were taken for analysis and plotting. Note that once the deflection exceeded 100 mm, the deflection was based on the actuator displacement. Table 4 presents the results of the tested diaphragm, while Figure 66 shows the load-deflection curves.

Table 4. Stiffness and strength properties of the tested diaphragms

Configuration	$K_{\Delta/1080}$ (kN/mm)	$K_{40\%}$ (kN/mm)	P_{peak} (kN)	Δ_{peak} (mm)
1	8.37	4.85	431*	-
2	6.35	3.19	395	152.7
3	5.85	2.98	257	121.2
4	4.33	2.30	197	130.4
5	5.14	2.11	112	179.6
6	14.26	14.34	389**	43.6
7	8.14	5.76	395	127.8

* The displacement measurement was unrecorded for this load, so the peak load does not appear on the graph

** The specimen was not brought to failure

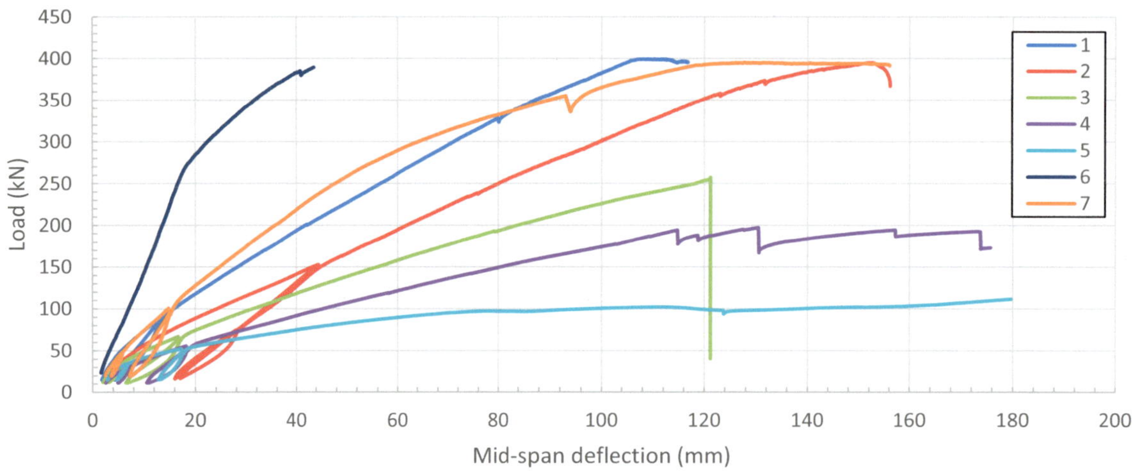

Figure 66. Load-deflection curves for the seven tested diaphragms.

Comparing configuration 2 to configuration 1 shows that the initial stiffness decreases by approximately 24%. Diaphragm configuration 2 represents a more realistic case, in which the glulam decking panels cannot run continuously between the shear walls.

Comparison of results from configuration 3 to those of configuration 1 (similar, but with thicker, 89-mm wood decking) shows that the initial stiffness and maximum load decreased by approximately 30% and 40%, respectively. This shows the impact of reducing the thickness of the glulam decking on the stiffness of the diaphragm.

Comparing configuration 4 and configuration 3 shows that the initial stiffness and maximum load capacity both decrease by approximately 25%. When compared to diaphragm configuration 2 (staggered but with thicker pieces of decking), the thickness reduction leads to a 30% lower initial stiffness and a 50% lower maximum loading capacity.

In configuration 5 and configuration 3, the glulam decking was the same, but the panel-to-panel joint ran parallel to the length for the former and perpendicular for the latter. A comparison of their respective results finds that the initial stiffness decrease by approximatively 12% and the maximum load by 57%.

Configuration 6, a decking made of CLT without discontinuities, exhibited the highest initial stiffness—more than 70% that of diaphragm configuration 1, despite the panel's thickness only increasing by 16 mm (an 18% increase). This important difference in stiffness may result from the in-plane mechanical properties of the continuous CLT panel (i.e., bending and shear moduli) governing the load-deformation behaviour. When there is discontinuity in the CLT panel, such as in diaphragm configuration 7, the initial stiffness decreases by about 43% as compared to configuration 6. This loss of rigidity shows the strong influence of the panel-to-panel assembly on the diaphragm's behaviour.

6.2 Nordic Tests at FPInnovations

In 2016, Nordic Structures contracted FPInnovations to perform in-plane bending tests on diaphragms with certain specifications. Nordic Structures has kindly allowed FPInnovations to share these test results for the purposes of this document. FPInnovations tested a total of three (3) diaphragms with different configurations. The test setup for the three configurations appears in Figure 67. The span and height of the diaphragm were 7315 mm (24 ft.) and 2034 mm (6 ft. 8 in.), respectively. The diaphragm was loaded with four (4) equally spaced loading points, excluding the supports (Figure 67).

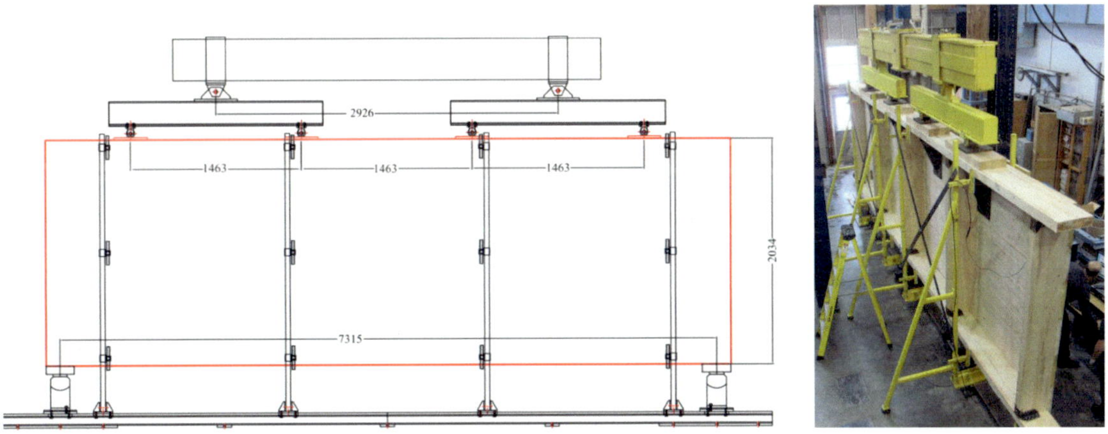

Figure 67. The test setup used for the Nordic diaphragms.

The tests used a hydraulic jack with a 500 kN capacity to achieve the ultimate loads for the diaphragm specimens. A computerized data acquisition system recorded load measurements at an acquisition frequency of 5 Hz. Other displacement measurement devices (Lasers and LVDTs) were in the positions shown in Figure 68, for a total of 10 channels.

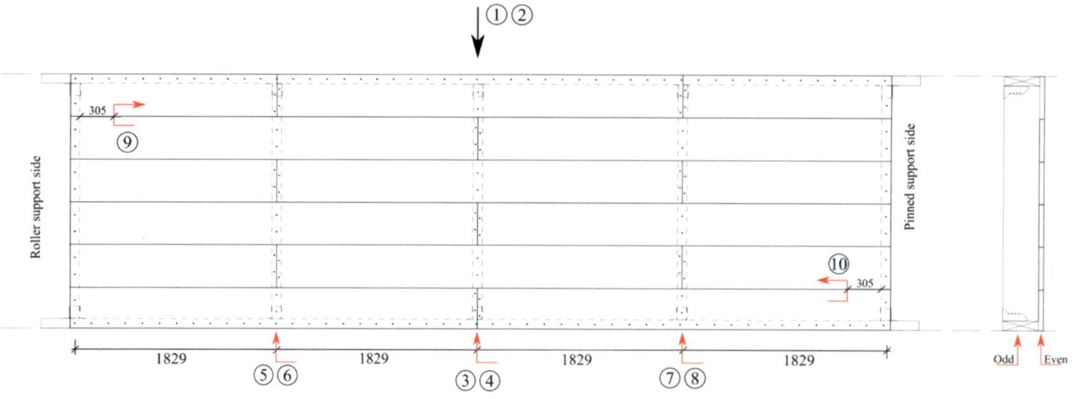

Figure 68. Diaphragm dimension with the position of the installed measuring devices.

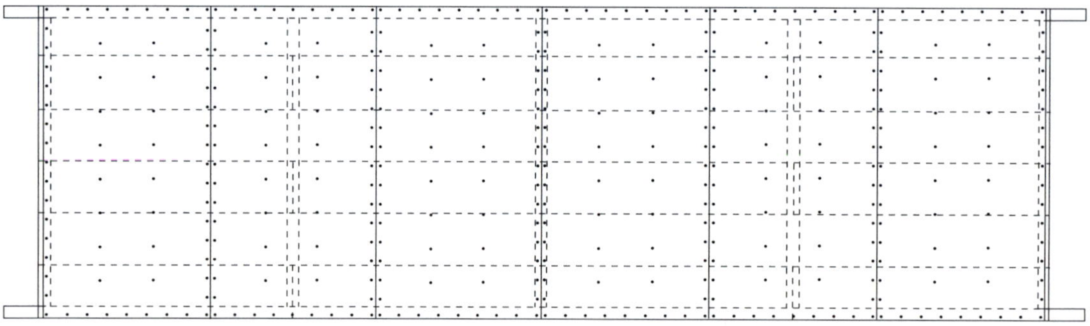

Figure 69. OSB panel fixed to the glulam decking for configurations 1 and 2.

Configuration 1 comprised a 44 mm–thick glulam decking with an 11 mm–thick OSB panel (Figure 69). The orientation of the glulam decking was perpendicular to the load. The glulam was nailed together with 2.5 x 44 mm nails installed at 30° with a spacing of 100 mm o/c, as shown in Figure 70. The OSB panel was nailed to the glulam with ϕ3.25 x 50.8 mm nails installed with a spacing of 150 mm on the perimeter of the panel and a spacing of 300 mm on its centre line.

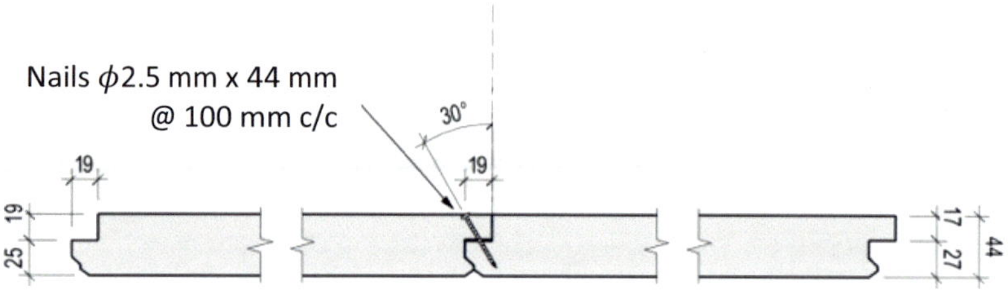

Figure 70. The nailed glulam decking of diaphragm configuration 1.

Configuration 2 was identical to configuration 1, except that the glulam decking elements were not nailed together. Configuration 3 comprised 89 mm–thick glulam decking without an OSB panel (Figure 71). The orientation of the glulam decking was parallel to the load, and the glulam elements were nailed together with 3.9 x 82.5 mm nails installed at 30° with a spacing of 100 mm.

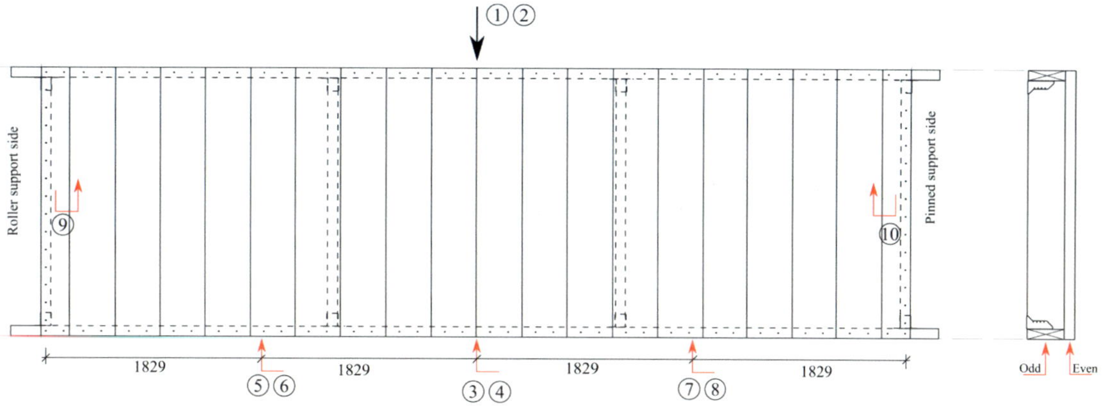

Figure 71. The nailed glulam decking of diaphragm configuration 3.

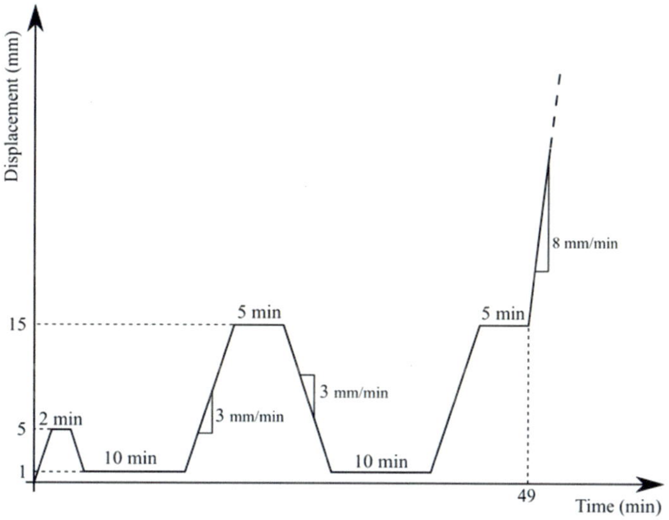

Figure 72. The loading protocol used to test all diaphragms.

The loading protocol used in the tests is shown in Figure 72. It involved a displacement of 5 mm, applied at the beginning of the test and maintained for two minutes. The displacement then decreased to 1 mm, which was maintained for 10 minutes. After that, the displacement increased to 15 mm and was maintained for 5 minutes, before being reduced to 1 mm and maintained for 10 minutes. Finally, the displacement increased to 15 mm and was maintained for 5 minutes. Thereafter, the displacement increased until failure. The test used a constant rate of 3 mm/min, except for the last increase, during which the rate was 8 mm/min.

6.2.1 *Results and Discussion*

Among the three configurations, diaphragm configuration 3 (glulam decking oriented parallel to the load) had the highest stiffness values for the mid-span deflection and the lowest resistance values. These results may seem abnormal, but most of the deformation was essentially due to shear deformation, and as is well known, elements parallel to the load take more shear stress. After the connection between the glulam decking elements yielded, it was difficult for these elements to take more shear stress because they were limited by the capacity of the connection. This explains the lower strength.

As expected, the stiffness and resistance results for diaphragm configuration 1 were higher than those for diaphragm configuration 2 because of the nails used in the glulam decking. These nails allowed an increase in stiffness of around 35% and an increase in resistance of around 20%.

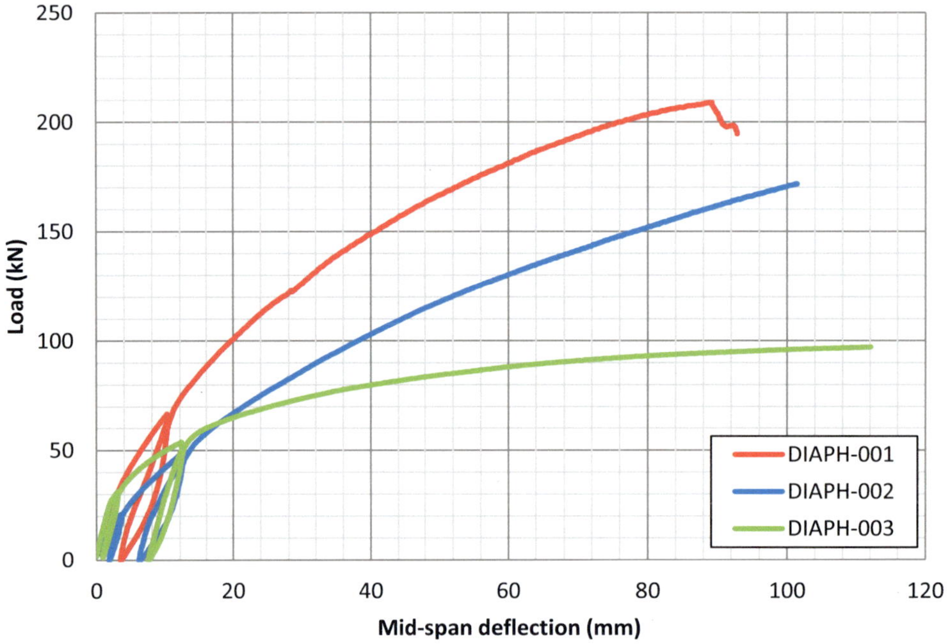

Figure 73. Load–mid-span deflection curves for the 3 diaphragms.

Figure 73 shows the mid- span deflection for each respective diaphragm configuration. Each curve is an average of the two laser measurements. Figure 74 shows the relationship between the load and relative slip for the glulam decking elements near the supports for each configuration. The numbers in parentheses in the legend represent the displacement transducer used, as per Figure 68 and Figure 71.

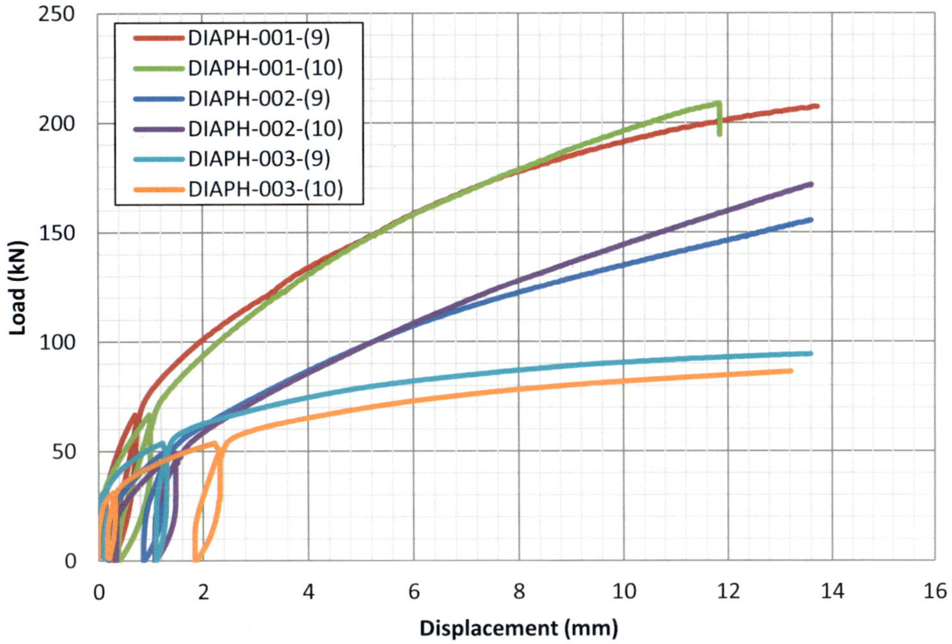

Figure 74. Load–Relative slip between the glulam decking near the supports, for each diaphragm configuration.

Tabulated stiffness and load values obtained from the tests appear in Table 5.

Table 5. Load, deflection, and stiffness values from the diaphragm tests

Configuration	$K_{\Delta/1080}$ [kN/mm]	$K_{40\%}$ [kN/mm]	P_{peak} [kN]	Δ_{peak} [mm]
1	7.44	5.59	208.8	89.1
2	4.52	3.20	171.6*	101.4*
3	6.10	3.43	97.1*	112.1*

* These values are the highest ones obtained during the testing and may not be the ultimate values

6.3 Research on Diaphragm Performance Conducted at UBC

Ashtari (2012) studied the in-plane behaviour of CLT floor diaphragms using detailed 2D numerical models developed in the finite element software ANSYS. She used a simple smeared CLT panel-to-panel connection model in the ANSYS floor models. The model ably captured the response of the panel-to-panel connections when compared to previously conducted connection tests. Although the number of floor diaphragms this study analyzed was limited, the results can illuminate some important aspects of the in-plane behaviour of CLT diaphragms.

The study found that the in-plane stiffness of a diaphragm depends on the in-plane "material properties," such as the mechanical properties of the CLT panels and the parameters defining the nonlinear response of the CLT panel-to-panel connections, as well as on the "geometrical parameters," such as floor diaphragm configuration, the number of connected CLT panels, and the dimensions of these panels. A parametric analysis examined the simple case of a diaphragm that consisted of nine CLT panels, each with an aspect ratio of approximatively 4.2 x 1 (Figure 75). The models of the connections between the panels were based on tested connections that used two pairs of STSs placed at 45° to the CLT face and 35° to the grain (connection configuration E). The continuous load was applied in the longitudinal direction of the CLT panels (perpendicular to the diaphragm length).

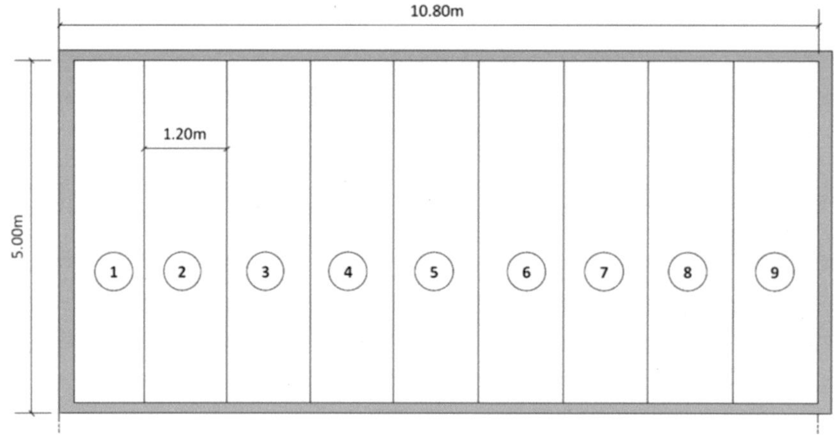

Figure 75. Diaphragm configuration (Ashtari, 2012).

The parametric analyses of the diaphragm showed that an increase in the initial stiffness of the panel-to-panel connections also increased the in-plane stiffness of the entire diaphragm. The curve in Figure 76 shows the relationship between the initial stiffness of the panel-to-panel connections and the diaphragm stiffness on a logarithmic scale, which is nonlinear. The stiffness $K_1 = 1.0$ is the average stiffness of the tested connection with two pairs of inclined screws, and was 18.74 kN/mm. The curve demonstrates that changes in diaphragm stiffness are larger when the initial connection stiffness is lower. For instance, if the initial connection stiffness increases by 10 and 1000 times from $K_1 = 1.0$, which is the tested connection, the in-plane stiffness of the CLT diaphragm increases by factors of 2.2 and 2.6, respectively. On the other hand, if the initial connection stiffness decreases 10 and 100 times from $K_1 = 1.0$, the in-plane diaphragm stiffness drops by factors of 5.8 and 75, respectively.

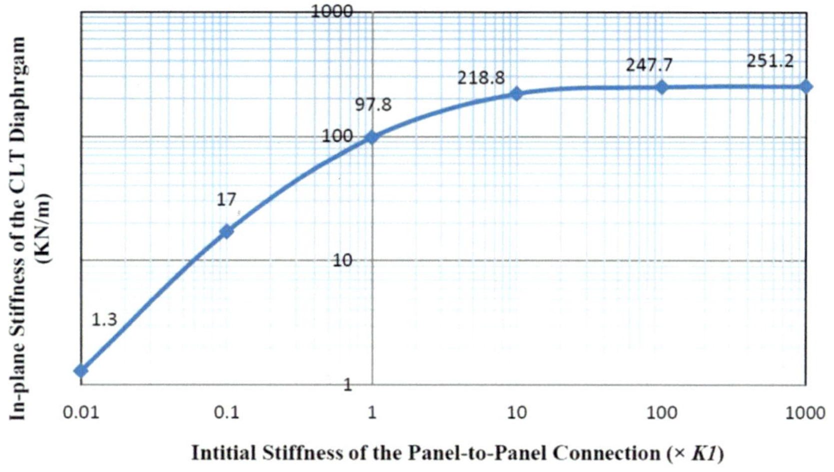

Figure 76. In-plane stiffness of a CLT diaphragm for different initial stiffnesses of panel-to-panel connections (Ashtari, 2012).

The analyses also showed that for softer panel-to-panel connections, most of the diaphragm deformation takes place in the connections, so the panel-to-panel connection behaviour controls the in-plane diaphragm behaviour. For stiffer panel-to-panel connections, however, most of the in-plane deformations occur in the CLT panels; therefore, the in-plane behaviour of the CLT diaphragm is controlled by the CLT panel properties. This would also explain the trend observed in Figure 76: the in-plane stiffness of the diaphragm changes much faster when the initial stiffness of the panel-to-panel connection is lower because the in-plane behaviour of the diaphragm depends on the connections. On the other hand, the incremental change in the in-plane stiffness of the diaphragm is much smaller for stiffer connections. This is because the in-plane stiffness of the CLT panels determines the in-plane behaviour of the diaphragm, so increases in the initial stiffness of the connection have only an insignificant effect on the stiffness of the diaphragm.

Ashtari (2012) also varied the number of CLT panels in a given diaphragm configuration, as shown in Figure 77, to determine the effect of the number of panels on the diaphragm stiffness. The number of panels varied from 3 to 9, using two different in-plane shear moduli for the CLT panels, representing significantly flexible panels (G_{xy} = 100 MPa) and regular stiffness (G_{xy} = 600 MPa), respectively. From Figure 77, it is evident that the in-plane stiffness decreased nonlinearly with an increase in the number of panels. The change is more significant with a smaller number of panels. Moreover, the in-plane stiffness of the CLT diaphragm with stiffer CLT panels changes more rapidly with the number of connected panels than do the more flexible ones.

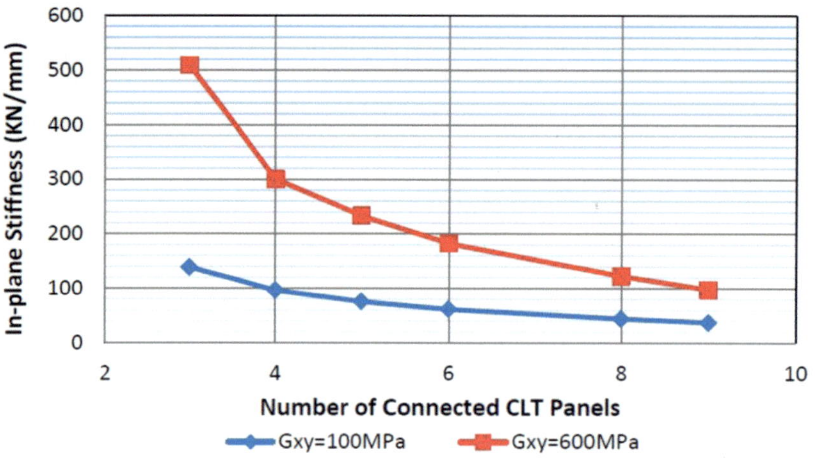

Figure 77. In-plane stiffness of a CLT diaphragm with different numbers of panels and two different in-plane shear moduli for the panels (Ashtari, 2012).

The explanation for this behaviour lies in the type of deformations that occur within the floor diaphragm. With fewer panels and thus a shorter length, the diaphragm acts as a deep short beam deforming almost entirely in shear. In this case, changing the in-plane shear modulus of the CLT panels affects the in-plane deformations quite significantly. With an increase in diaphragm length (more CLT panels and panel-to-panel connections), the role of connections in the total in-plane deformation of the CLT diaphragm becomes more significant. In other words, the change in the shear modulus of the CLT panels has less effect on the in-plane stiffness of the CLT diaphragms with a higher number of connected CLT panels.

6.4 Analytical Research on the In-Plane Flexibility of CLT Diaphragms

The Kore University of Enna in Italy, the Trees and Timber Institute of Italy, and the University of Edinburgh in Scotland have conducted a collaborative study on the flexibility of CLT diaphragms (D'Arenzo et al., 2019). As the strength and deformation properties of diaphragms strongly depend on the connection behaviour, the first part of this study included 12 tests on diaphragm connections. Tested connections used 85 mm–thick CLT panels made up of five 17 mm layers with lap joints that were 50 mm wide. They used Simpson Strong-Tie Eurotec 8.0 x 80 screws, with an inner thread diameter of 6 mm and an outer thread diameter of 8 mm. The mean moisture content of each specimen was 10.6%, with a coefficient of variation of 8.5%. The mean density, corrected for 12% moisture content, was 448 kg/m^3 with a coefficient of variation of 1.5%. In addition to these tests, the analytical part of the study used results from other tests in the literature.

The analytical part of the study also developed two models, a plane model and an equivalent frame model. The plane model used 2D membrane elements for the CLT panels. The modulus of elasticity of the CLT in the longitudinal (L) and transversal (T) direction were determined using the composite theory. An equivalent shear modulus that takes into account the number of layers, the total thickness of the panels and the width of boards was used in the modeling, as per Brandner et al. (2017). An example of the plane model appears in Figure 78. As shown, the CLT floor comprises of *m* panels with length *b*, subjected to an in-plane uniform horizontal load q. The floor span is *L* and total width of the floor is *B*, which in turn is equal to *b* x *m*. The CLT panels have an odd number of orthogonal layers with a total thickness equal to *t*.

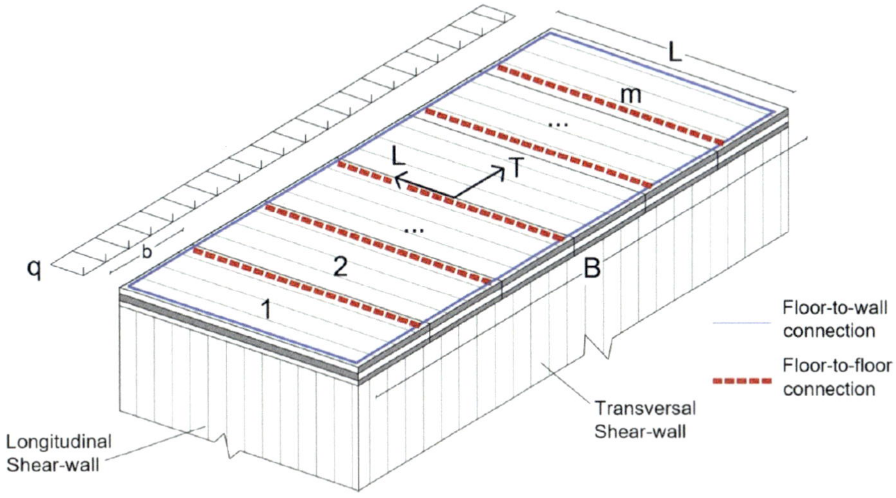

Figure 78. An example of the plane model (D'Arenzo et al., 2019).

The connections between adjacent floor panels (in L-direction) were modeled by line springs with an elastic stiffness per unit length equal o $K_{f\text{-}f,L}$, while connections transverse to the panels (in the T direction) were modeled with a stiffness per unit length equal to $K_{f\text{-}f;T}$. For the connections between the floors and the walls, the study used line springs acting only in the direction parallel to the floor edges, with stiffness per unit length equal to $K_{f\text{-}w}$. The stiffness of the floor-to-wall connections in the orthogonal direction was assumed to be equal to zero since the design of the vertical supporting elements is not generally designed to withstand any out-of-plane lateral loads.

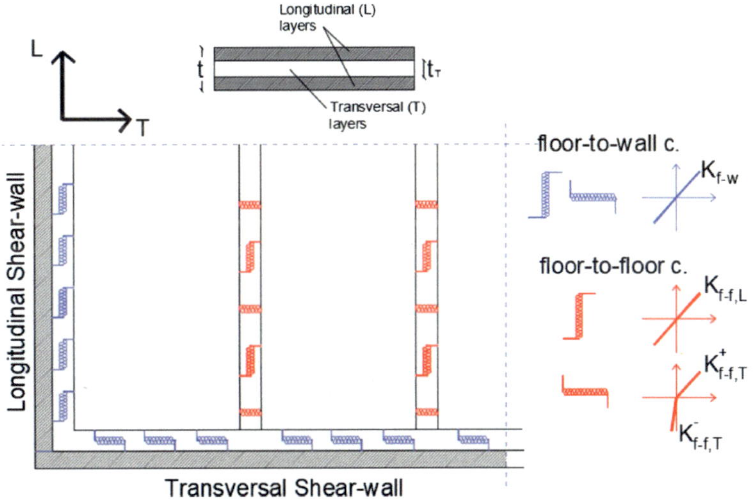

Figure 79. Mechanical properties of connections in the plane model (D'Arenzo et al., 2019).

The study assumed the behaviour of the CLT panels and connections would be linear elastic, except for the floor-to-floor connection in the T direction. To take into account the interaction between two adjacent floor panels, it used different values of stiffness $K_{f\text{-}f,T}$ when line springs work in tension or compression. When two adjacent panels move away from each other, the stiffness of the spring correlates to that of the floor-to-floor connection, whereas when the panels come in contact with each other, a rigid contact element was adopted. As a result, the mechanical behaviour of the floor-to-floor connection in the T direction was modeled by elastic non-linear links with linear elastic behaviour in tension $K^{+}_{f\text{-}f,T}$ in tension and rigid behaviour $K^{-}_{f\text{-}f,T}$ in compression. The mechanical properties of the connections of the plane model are shown in Figure 79.

To facilitate the representation of the results of sensitivity analyses, the study defined a spring model for the in-plane behaviour of CLT diaphragms. Each deformation contribution (i)–(vi) is represented by an equivalent horizontal spring, working either in parallel or in series (Figure 80). If we assume that the diaphragm acts as deep beam, we can then split the total deformation into mainly shear and bending contributions. The diaphragm displacement due to shear is related to (i) the in-plane shear deformation of the panels, which depends on the equivalent shear modulus and the total thickness t of the panel; (ii) the contribution of the floor-to-floor connection along the L direction between the panels; and (iii) the contribution of the floor-to-wall connections along the longitudinal walls. The displacement due to bending depends on (iv) the in-plane bending stiffness of the CLT panels, which is influenced by the equivalent modulus of elasticity and the total thickness of the panel; (v) the stiffness of the floor-to-floor connections along the T direction; and (vi) the stiffness of the floor-to-wall connections along the transversal shear walls.

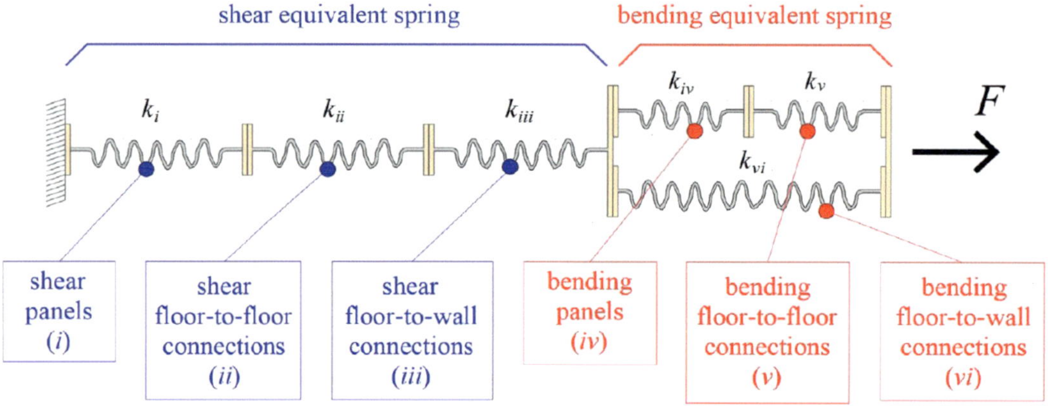

Figure 80. Spring model for the in-plane behaviour of CLT diaphragms (D'Arenzo et al., 2019).

A simplified equivalent frame model provided sensitivity and parametric analyses for the diaphragms, which were then verified against the plane model. In this way, the study took advantage of the low computational cost of the equivalent frame model to conduct an extensive sensitivity analysis to identify the most important deformation mechanisms contributing to in-plane deformation in CLT floor diaphragms.

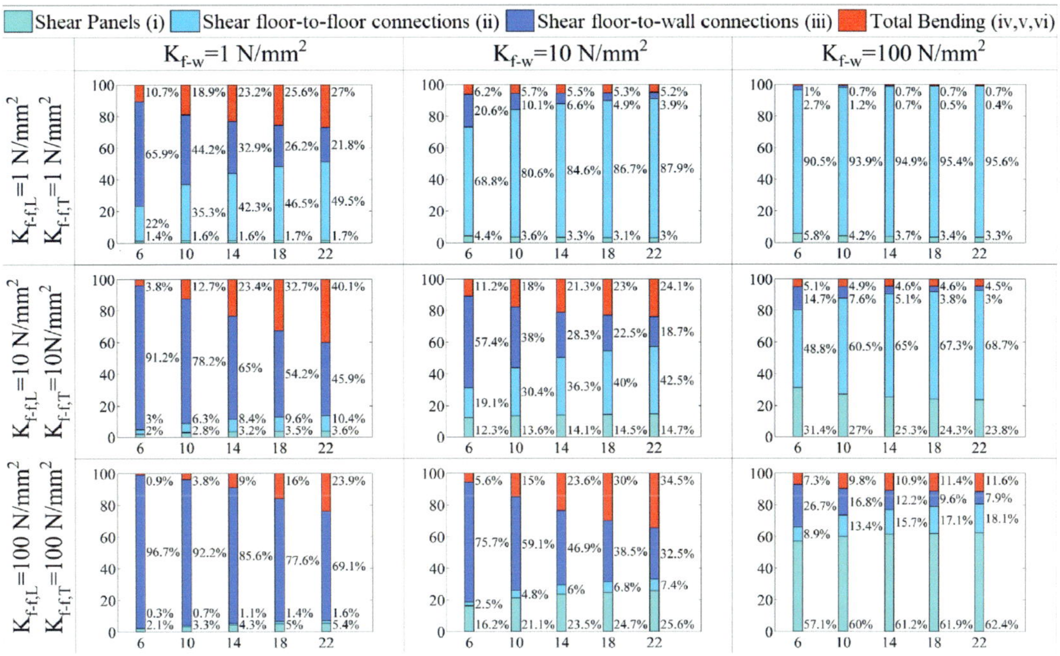

Figure 81. Shear and global bending deformation contributions [in %] for different diaphragm widths B [in m] (D'Arenzo et al., 2019).

Analyses showed that in most cases, shear deformation was the main contributor to the total displacements of the diaphragms (Figure 81). It ranged between 59.9% and 99.3%, which was most probably due to the low aspect ratio of the diaphragms, which ranged between 1.2 and 4.4.

The results show that the connections between the CLT diaphragm panels to the walls underneath substantially affect diaphragm deformation (flexibility). The connections between the diaphragms and the walls tend to redistribute the bending moment but do not significantly affect the shear deformations between panels. The equivalent frame model captured this effect, while a simply supported beam model did not.

Within the limits of realistic connection stiffness, material properties, and geometry, the analyses determined that the stiffness of the diaphragm was not sensitive to the thickness of the floor panels. The floor geometry and connection stiffness were found to be the key influencing parameters. As the aspect ratio of the diaphragms in this study varied between 1.2 and 4.4, the flexibility of the floor diaphragm was generally dominated by slip between panels, rather than by panel rotation or global in-plane bending. The flexibility thus depends on the stiffness of the connections under this type of load.

An analytical study by D'Arenzo et al. (2021) investigated the in-plane behaviour of CLT diaphragms to determine in which conditions one can treat the diaphragm as rigid or flexible. The authors used the criteria for a flexible diaphragm as defined in Eurocode 8 (CEN, 2004b) and ASCE 41 (American Society of Civil Engineers [ASCE], 2017), which differ significantly. The criteria in these standards can be expressed as $\Delta_{floor}/\Delta_{wall} \leq \alpha$, where Δ_{floor} and Δ_{wall} are the displacements of the floor and the wall, respectively, and α is the coefficient for a "rigid diaphragm condition." This coefficient is equal to 0.1 in Eurocode 8 and 0.5 in ASCE-41. As discussed in the previously cited paper (D'Arenzo et al., 2019), the wall-to-floor connections also contribute to the diaphragm deflection; when that influence is added to the wall displacement, the relationship becomes $\Delta_{floor}/\Delta_{sup} \leq \alpha$, where Δ_{sup} is the deflection of the diaphragm support; the latter can be expressed as $\Delta_{sup} = \Delta_{floor} + \Delta_{f\text{-}w}$, where $\Delta_{f\text{-}w}$ is the deflection due to wall-to-floor connections. Taking the floor-to-wall connection deformation into consideration, the authors introduced a new condition, factor β, for structures with flexible wall-to-floor connections, such that $\Delta_{f\text{-}w}/\ \Delta_{floor} \leq \beta$.

As shown in Figure 78, the study team conducted a parametric analysis on a single-storey CLT building, varying its geometry and connections to investigate different case scenarios. Linear elastic analyses using SAP2000 software showed, as expected, that in-plane behaviour varies in relation to the lateral stiffness of the shear walls and the stiffness per unit of length of the floor connections. The in-plane flexibility of the diaphragms was investigated with a focus on the ratio between the floor and the vertical support displacement, $\Delta_{floor}/\Delta_{sup}$, and the distribution of the lateral load between the shear walls. The numerical analyses in this study showed that for the CLT floors to have rigid diaphragm behaviour, the ratio $\Delta_{floor}/\Delta_{sup}$ should be less than or equal to 0.3 ($\Delta_{floor}/\Delta_{sup} = \alpha \leq 0.35$) for a span (distance) between the shear walls $A \leq 7$ m. Also, the β factor should be limited to 0.25 for a rigid diaphragm condition. Note that these results arise from an analysis of a single building configuration. Further investigation is needed to include different CLT building archetypes and confirm these results using linear and nonlinear dynamic analysis.

6.5 Testing of CLT Diaphragms in the U.S.

6.5.1 Testing at Colorado State University

To determine the performance of CLT diaphragms, a study at Colorado State University conducted four tests on two different configurations (Kode et al., 2021). All diaphragms were made of E1 grade CLT panels that used 1950f-1.7E machine stress rated spruce-pine-fir lumber in all parallel layers and No. 3 spruce-pine-fir lumber in all perpendicular layers (ANSI/APA, 2019). The CLT diaphragms consisted of eight 2.14 × 1.14 × 169 mm panels and two smaller pieces, 2.14 × 0.42 × 169 mm, at the base of the cantilever diaphragm, making the total size of the diaphragm 5 × 4.27 m (16 ft. 5 in × 14 ft.). Chord members consisted of 89 × 190 mm (4 × 8) rough sawn timber. The tests used an actuator connected at the free end of the CLT diaphragm to a loader bar comprising a steel plate and two pieces of steel tubes. The steel plate attached to the actuator head, which then connected to the two pieces of tube steel that ran along the top and bottom of the CLT diaphragm's free end. At the base of the diaphragm (the fixed end), the sides of the CLT panels were attached to a W 6 × 14 steel beam with 54 lag screws (7.9 x 254 mm).

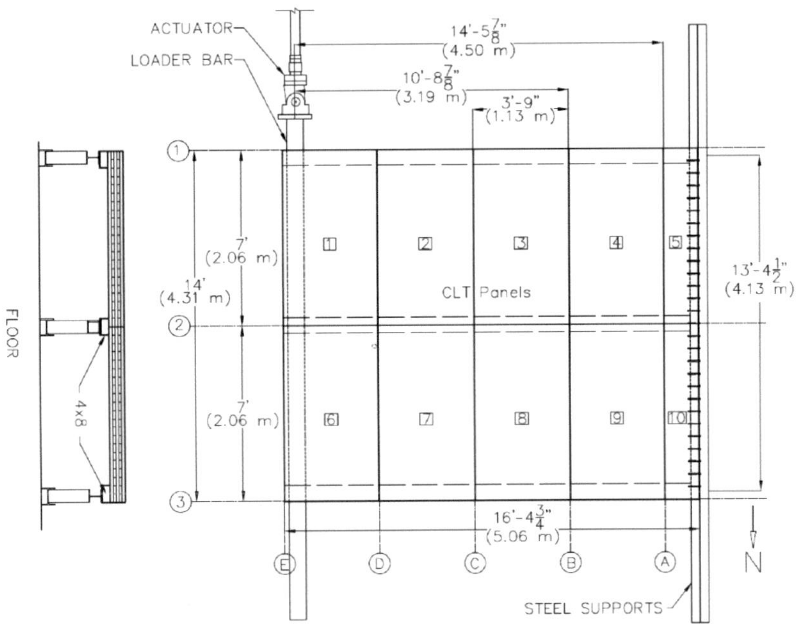

Figure 82. A plan of the test setup (Kode et al., 2021).

Diaphragm configuration 1 had a low-strength tension chord connection to the support, while that in configuration 2 used a higher strength one. The weaker cord connection used 228.6 × 152.4 × 3.5 mm 10-gauge steel plate, with eight 16d box nails (3.4 × 88.9 mm) along Lines A, B, C, and D (Figure 82). In the direction along Lines 1, 2, and 3, CLT panels were connected for shear to supporting framing members with two rows of eight 6.35 x 203 mm STSs spaced at 152 mm on centre on each side of a panel joint (16 screws along each line), except at CLT Panels 5 and 10, which used four similar STSs. Configuration 2 strengthened chord-to-support connections by adding a Simpson Strong-Tie HDU8 hold-down and 10-gauge steel connectors with 32 evenly spaced 6.4 x 152 mm STSs and 16 screws on each side of a joint, in addition to the four screws connecting the chord to panels 5 and 10 (Figure 82). The Simpson Strong-Tie HDU connections spanned the undersides of Panels 4 and 5 (9 and 10) with eight screws in each panel, while the 10-gauge steel tension connector spanned the same panels but on the upper panel surface (Kode et al. 2021).

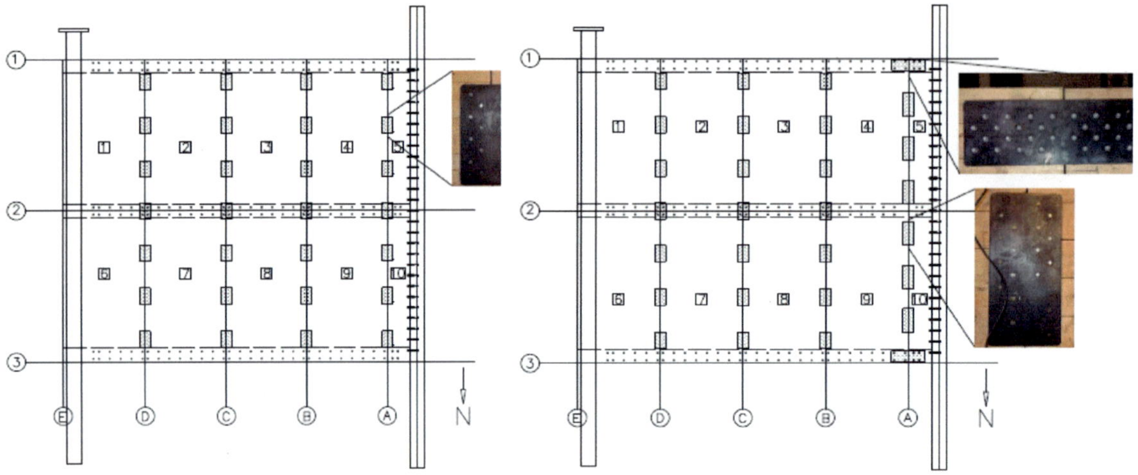

Figure 83. Connector layouts for configuration 1 with a weak tension chord connection (left) and configuration 2 with a stronger tension chord connection (right) (Kode et al., 2021).

Testing occurred under displacement control using the CUREE loading protocol. The maximum displacements and loads obtained during all tests appear in the table in Figure 84.

Diaphragm size	CLT grade	CLT panel thickness [mm (in.)]	Configuration	Test No.	Reference displacement [mm (in.)]	Max displacement (mm)		Max force (kN)	
						Positive excursion	Negative excursion	Positive excursion	Negative excursion
5 × 4.27 m (16 ft 5 in. × 14 ft)	E1	169 (6.65)	1	1	12.7 (0.5)	18.9	18.9	30.2	31.7
				2	25.4 (1)	37.81	37.55	42.1	60.4
				3	152.4 (6)	105.9	107.8	67.88	67.61
			2	4	76.2 (3)	188.8	150.7	179.2	174.2

Figure 84. The result summary of the four tests (Kode et al., 2021).

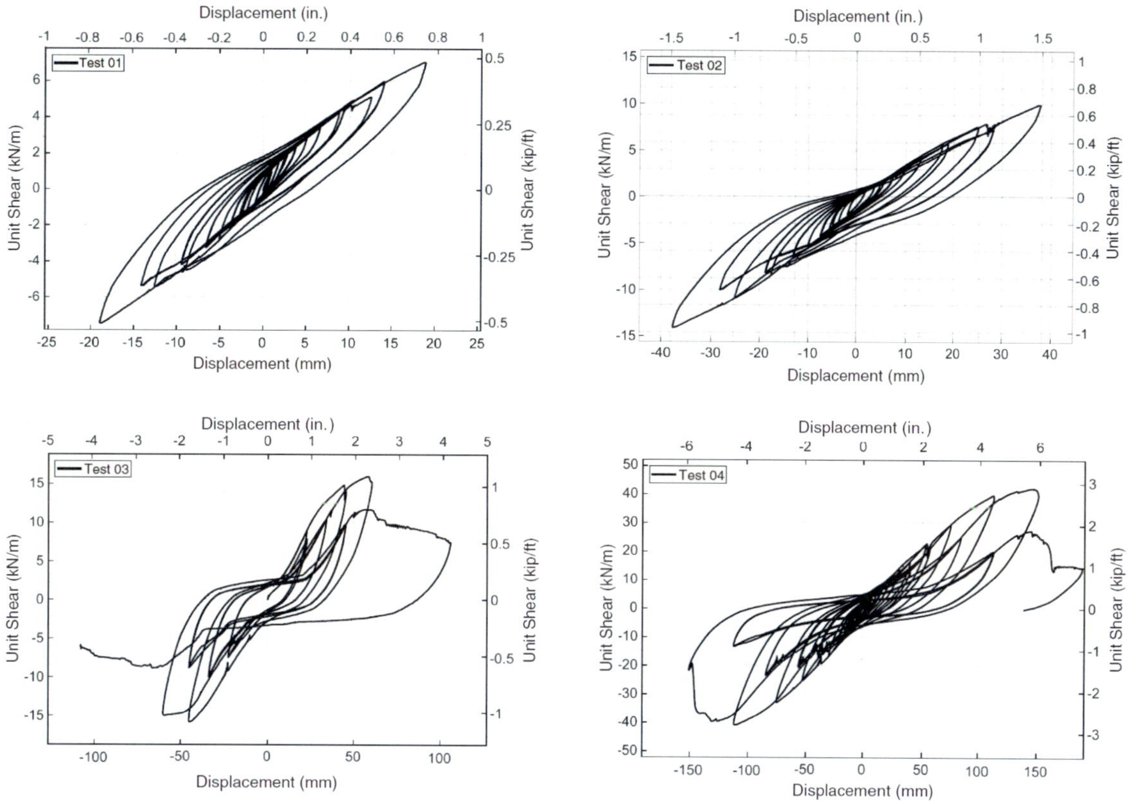

Figure 85. Load-displacement hysteretic curves from the four tests; configuration 1 (top and bottom left) and configuration 2 (bottom right) (Kode et al., 2021).

During tests 1 and 2 of configuration 1 (Figure 85 top), the diaphragm remained in the elastic range. There was no visible structural damage to the CLT panels or the connections in the diaphragm, so no repairs took place after these tests. During test 3 for configuration 1, the low-strength chord connection failed in tension. Test 4 of configuration 2 showed that a higher strength chord connection enabled the development of the full design strength of the diaphragm shear connectors (Figure 85 bottom). These tests also confirmed the rigid behaviour of the CLT panels and the energy dissipation mechanisms in the connectors. The tests indicate that the chord requires careful consideration in the design of CLT diaphragms to enable the shear connectors to develop their full strength.

6.5.2 American Wood Council Diaphragm Tests

The diaphragm tests by Kode et al. (2021) that are the subject of the previous section, along with all other tests on mass timber diaphragms, have shown that the panels themselves behave rigidly, with most energy dissipation occurring in the connections. The tests also highlighted the importance of designing chords with sufficient strength. Building on the work of Kode et al. (2021), the American Wood Council (AWC) conducted tests on two full-scale CLT diaphragms with different configurations at the Weyerhaeuser Laboratory in Federal Way, Washington (Line et al., 2022). These tests represent the first controlled experimental evaluation of the Special Design Provisions for Wind and Seismic (SDPWS) design requirements for CLT diaphragms. The primary objective of these tests was to verify that CLT diaphragms in accordance with the SDPWS design provisions could develop the calculated nominal capacities, which represent a minimum level of overstrength that is 2.8 times the allowable stress design seismic design shear capacity.

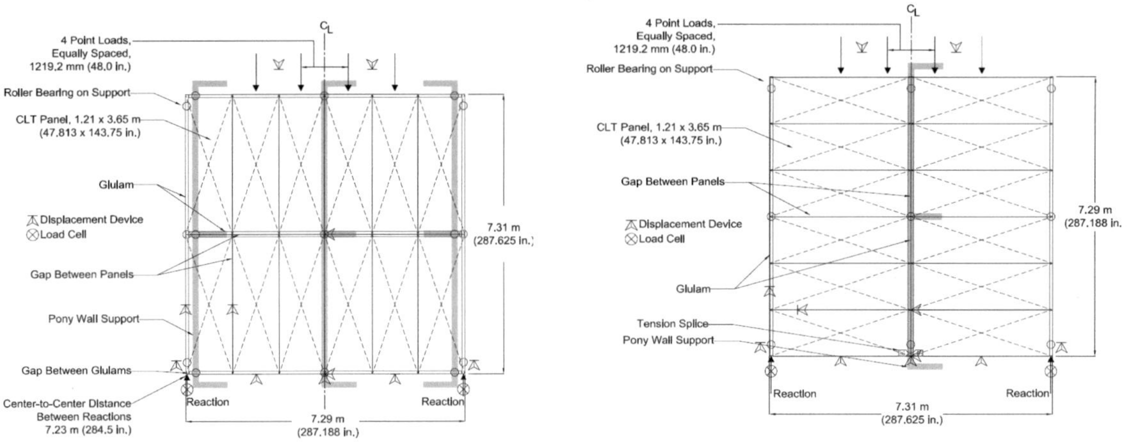

Figure 86. Diaphragm assemblies for diaphragm CLT 01, with the long dimension of the CLT panels oriented parallel to the load (left) and CLT 02, where they are perpendicular to the load (right) (Line et al., 2022).

The diaphragms were made of 1.22 × 3.66 m (4 × 12 ft.) CLT panels with the panel ends connected to a common glulam beam framing, while plywood splines and fasteners served as connections between the CLT panels. The study tested one diaphragm with the CLT panels oriented parallel to the load (CLT 01), while the second had the CLT panels oriented perpendicular to load (CLT 02) (Figure 86). The panels were 3-ply 105 mm thick and made of spruce pine, proprietary grade E1M5. 24F-V8 Douglas fir glulam beams 228.6 mm (9 in.) deep formed the gravity load frame used to support the CLT panel weight. The width of the beam was 79.4 mm at the diaphragm perimeter and 130.2 mm at the central bearing location. The panel-to-panel spline connection used 18.3 mm (23/32 in.) plywood connected to the CLT panel with 8d common (3.3 × 63.5 mm) power-driven nails with 7.1 mm head diameter. For the CLT 01 diaphragm, the glulam beams at the diaphragm edges served as tension and compression chords. For CLT 02, the edge panels themselves were designed as the tension and compression chords, with a custom 4.8 mm thick metal tension strap tying the panels together at the tension chord. The 7.32 × 7.32 m (24 × 24 ft.) diaphragms were tested under monotonic loads in general accordance with the ASTM E455 standard, with four loading points.

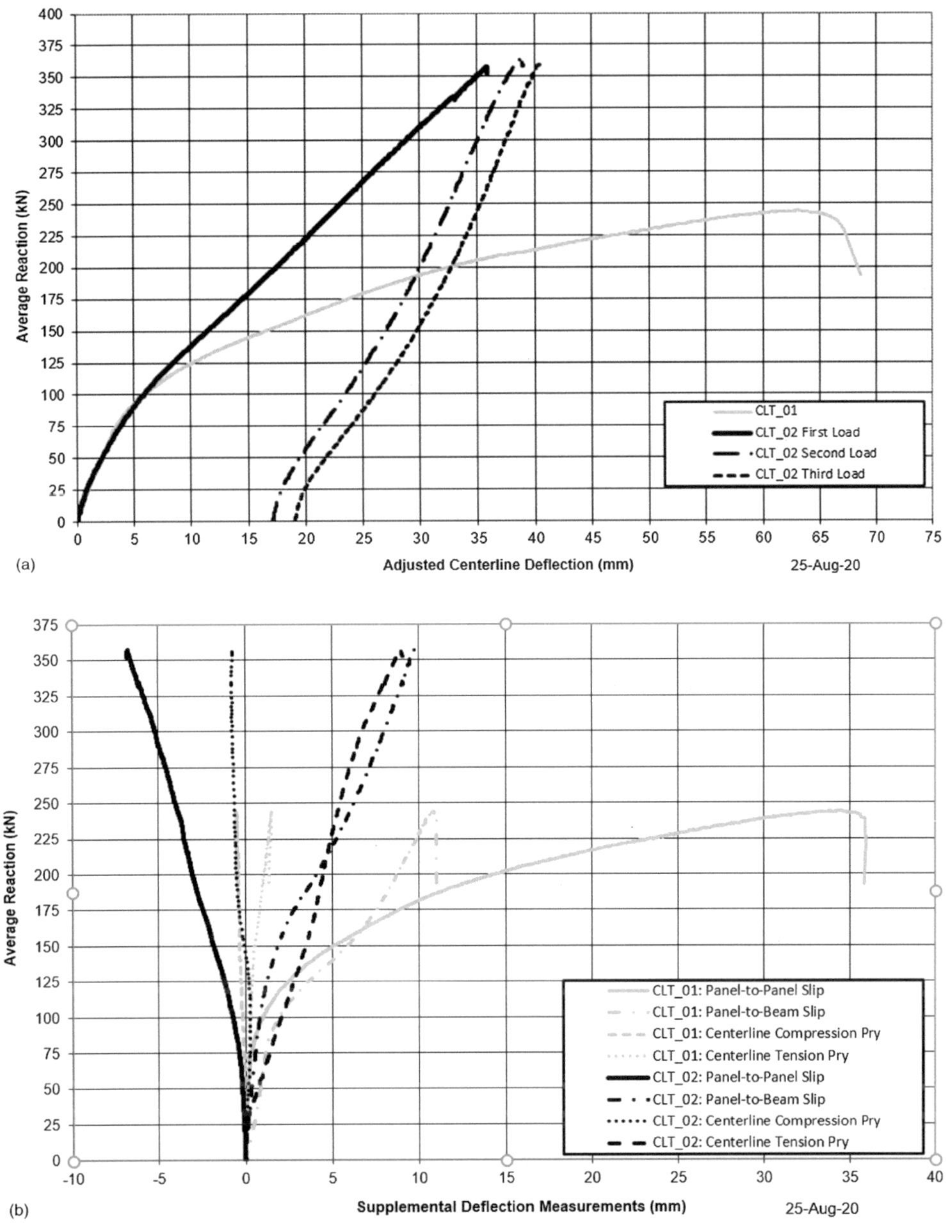

Figure 87. Load-deformation curves for both diaphragms: (a) average reaction versus centreline deflection; and (b) connection deformations (Line et al., 2022).

Both diaphragms in this study met the intended strength performance and developed their calculated nominal shear capacity of 170 kN (38.1 kips) per reaction. CLT 01 failed at a load that was 44% greater than the calculated nominal capacity. CLT 02 did not fail at the maximum applied load, which was 115% greater than the calculated nominal capacity. These findings confirmed that the design procedure provided sufficient overstrength to achieve the intended load capacity. Neither of the tested diaphragms experienced brittle wood or metal member failures. The study also induced ductile diaphragm deformation mechanisms at the panel-to-panel spline, panel-to-beam, and tension strap (CLT-02 only) connections. As shown in Figure 87b, connection mechanisms within each diaphragm provided a nonlinear response beyond the proportional limit, which was subsequently reflected in the overall diaphragm deformations shown in Figure 87a. Diaphragm CLT 01 achieved a deformation at its ultimate load that was more than 16 times the measured deformation at the LRFD design load level. Even though CLT 02 was not pushed to failure and thus had an artificially truncated response, the measured peak load deformation still occurred past the proportional limit load, at 8.5 times the deflection measured at the LRFD design load. Diaphragm CLT 01 was the only one where failure was achieved. It failed at the panel-to-panel nailed spline connection, as expected. Ultimate failure was related to the withdrawal of the spline nails from the CLT panels. Load-deformation behaviour was dominated by connection slip, with the panels themselves rotating as rigid bodies.

The authors noticed that with the high load capacities achieved, there is the potential to either take advantage of this additional strength in the design or to make the design requirements less conservative. Some alternative load paths and sources of additional strength were apparent during the testing that had not been specifically accounted for in the design process.

6.5.3 Shaking Table Tests on Mass Timber Diaphragms

Barbosa et al. (2019) designed a test structure and conducted research (Barbosa et al. 2021) to establish a new understanding of CLT diaphragm responses during intense ground motions. The mass timber structure shown in Figure 88 was constructed on the University of California San Diego shaking table to evaluate the seismic performance of two floor systems and three interchangeable shear wall systems (Pei et al. 2019; van de Lindt et al. 2019).

Figure 88. Two-story mass timber structure tested during the summer of 2017 at the University of California San Diego outdoor shaking table (Barbosa et al., 2021).

The two diaphragm configurations remained unchanged throughout the testing program. The diaphragm on the first level was connected with plywood surface splines and partially threaded STSs. The roof was a CLT-concrete composite system, in which the CLT panels were interconnected using surface spline connections. The CLT panels connected to a concrete layer using inclined screws, which allowed for larger floor spans on the roof. Data from the testing, including accelerations measured in different locations in the diaphragms, served to assess the influence of distinct lateral-resisting systems on mean floor accelerations. The data also helped study the transfer of inertial forces within the diaphragms. In addition, the relative displacements measured between CLT panels, as well as between CLT floor panels and glulam beams, provided valuable information on the stiffness and capacity of the connections.

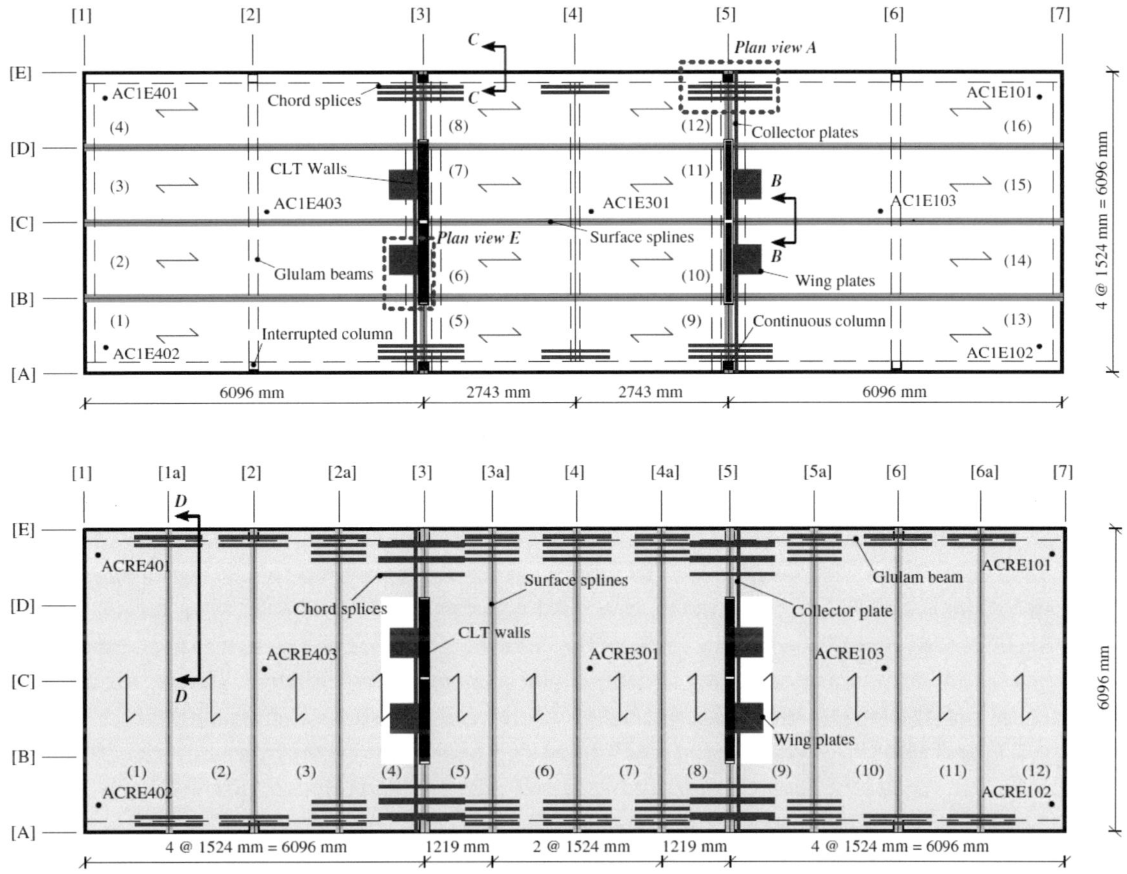

Figure 89. Structural layouts for phase 1 of testing: floor level (top); roof level (bottom) (Barbosa et al., 2021).

The diaphragms sustained a total of 34 shaking table tests during the three testing phases. Each phase used a different LLRS: Post-tensioned rocking walls in phase 1, nontensioned innovative rocking walls in phase 2, and a platform construction system with CLT walls connected to the diaphragms through shear anchors and rod hold-downs in phase 3. Results indicated that the alternative design method in ASCE 7-16 (Ghosh, 2016) provided a reasonable distribution of accelerations along the structure height when using conventional platform construction systems. Although the method was conservative, it failed to predict the floor accelerations for the CLT rocking wall systems the study tested. Despite the increases in ground motion intensity from Design-Basis Earthquake level to Maximum Credible Earthquake level, in phases 1 and 2, the LLRSs limited floor accelerations because the rocking lengthened the fundamental period of lateral vibrations and transmitted less force than shear walls cantilevered rigidly from the foundation.

The design procedure behind this work considered the diaphragms as deep beams, neglecting the additional flexibility due to the splines located between CLT floor panels. The in-plane distribution of accelerations showed that most of the diaphragm response was characterized by larger accelerations at the diaphragm (cantilever) corners, while mainly uniform acceleration was observed at the central core of the diaphragm. Relative displacements during the shaking table tests revealed that the connections responded mainly in the elastic range, as intended by the design objectives. The design was successful, given that all test phases saw little to no damage. There were smaller deformations for the slip movements than for the separation motions in the splines: these were limited to 1.0 and 2.5 mm, respectively. Based on test results on splines available in the literature (Closen, 2017; Taylor et al., 2020), the conclusion was that the spline connections in the building diaphragms exhibited linear elastic behaviour. The design of the chord splices neglected the contribution of the surface splines, resulting in an overestimation of design chord strains and forces that is conservative in design but requires careful assessment to ensure ductile modes of failure in the diaphragms. Finally, the rocking of the walls at the base and between the floor and wall may induce uplift between adjacent panels at the diaphragm level and plans for displacement compatibility in future designs should consider this mode of deformation, especially for platform construction.

6.6 Testing and Numerical Studies on Hybrid Mass Timber Diaphragms

Hybrid structural systems that combine wood with other materials provide engineers with a wide range of solutions to satisfying code requirements. Hybrid wood-based systems include many construction methods and different levels of integration of materials and components. Loss et al. (2018) investigated an innovative hybrid timber-steel solution for floor diaphragms (Figure 90b) that they developed by joining CLT panels (Figure 90a) with cold-formed customized steel beams. Each floor module then connects to the primary beams with ad hoc steel links (Figure 90c) manufactured by welding steel plates to one side of 350 mm–long steel pipes. To assemble the prefabricated components, one need only fasten the steel links using torque-controlled tightening bolts and insert STSs at the edge sides of the CLT panels. These connections also allow for rapid later floor replacement and accommodate any structural misalignments or imperfections (Loss et al., 2018). The fundamental repeatable floor unit was constructed off-site under controlled environmental conditions by joining a CLT panel with two custom-shaped cold-formed steel beams to provide composite action (Figure 90d). Figure 90 (bottom) shows the details of the customized thin-walled section cold-formed steel beams developed for this hybrid floor to maximize strength and save materials.

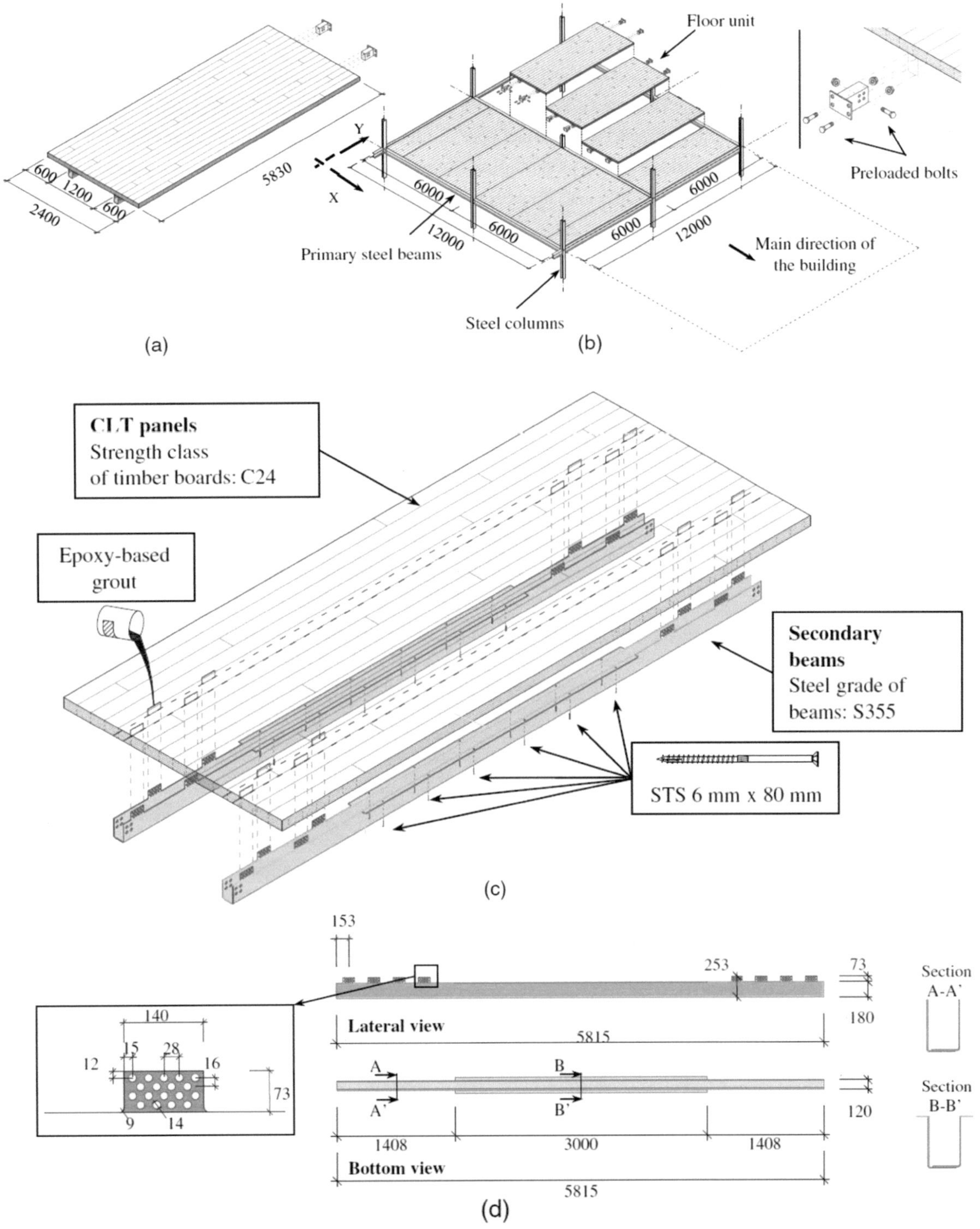

Figure 90. Hybrid floor system: (a) individual segment; (b) floor assembly; (c) beam-to-beam connections; (d) manufacturing process for the floor (Loss et al., 2018). All dimensions in mm

The objective of the research was to characterize the in-plane behaviour of this innovative hybrid steel-CLT floor system using analytical and experimental work. The work investigated the effects of reinforcing the steel profiles, increasing the stiffness of the connections, and varying the geometry of the basic components. The study also considered the effects of the in-plane response of the floors on the shear wall load distribution and the inter-storey drift of the building, varying the floor aspect ratio, shear wall stiffness, and layout. It applied a hybrid numerical simulation approach that performed each analysis using two numerical models in series. The first model, referred to as the micromodel, was developed in Abaqus to describe the local behaviour of several key elements, such as the flange yielding of steel links and transversal deformation of the steel profiles. The second model, referred to as the macro model, the entire floor was modeled using SAP2000, considering the responses of elements characterized in the micromodel. The models were verified against Loss and Frangi's (2017) experimental program, which included three tests on a full-size hybrid floor specimen.

Results showed that independent of the LLRS, the layout of the floor's subcomponents plays an important role in diaphragm behaviour. Depending mainly on the offset of the secondary steel beams as compared to the CLT panel edges and the thickness of the CLT panels, the in-plane stiffness of the diaphragm can increase from 21 to 200 kN/mm, while the yield capacity can increase from 268 to 852 kN. Joints that connect each modular floor unit to the primary steel beams have a significant impact on the diaphragm behaviour, bounding its elastic range. The diaphragm stiffness and its plastic deformation can be adjusted by designing different joint details.

As expected, the location of the restraints and the load distribution affected the behaviour of the diaphragm. The use of realistic load patterns is highly recommended. When the restraint arrangements reduce the efficiency of the diaphragm, one can restore its performance by reinforcing the cross sections of the steel members at their ends (e.g., using welded plates). Placing at least a pair of walls below the middle of the diaphragm and symmetrically arranging the other walls starting from the perimeter leads to a more balanced load distribution. Independent of the relative wall-to-diaphragm stiffness, the difference in the force acting on the internal and external walls (LLRSs) in this study was always less than 17%.

Assuming the same floor area and number of shear walls, a square floor plan offers considerable advantages compared with wider floor plans in terms of global lateral stiffness, ensuring that the load at each story is distributed through the bracing elements without a disproportionate horizontal deformation of the whole system. If it is necessary to limit lateral drift, the best option is to use sufficiently stiff vertical bracing elements, which will directly impact the diaphragm design. The ratio of the wall and diaphragm stiffness ($K_{wall}/K_{diaphragm}$) should be at least 0.5 to obtain a quasi-uniform (difference in forces less than 10%) distribution of the load within the walls. The common practice of designing shear walls using the envelope approach (based on the largest force among the flexible and rigid diaphragms assumption) proves not to be a rational way to estimate the corresponding level of diaphragm drift used to verify the damage limit state of the building.

7 DESIGN OF MASS TIMBER DIAPHRAGMS

Codes and standards around the world offer limited guidance on the design of mass timber diaphragms. This section presents the information provided in Canadian, U.S., and other codes and material standards. It also explores some of the guidelines that currently exist in this area. The analysis of a given diaphragm should make use of some of the methods discussed in section 3 of this document.

7.1 Diaphragm In-Plane Flexibility

Floor and roof diaphragms in buildings carry the vertical dead and live loads, but they also transfer the lateral loads imposed by wind and seismic actions to the vertical components of the LLRS below. In the latter case, the diaphragms rely on their in-plane strength and stiffness to transfer the imposed loads. In multi-storey buildings, where the diaphragms comprise reinforced concrete slabs or steel decks with structural concrete topping, the diaphragms have quite large in-plane stiffness, and they act as rigid bodies. In wood-frame structures, the situation is often the opposite because the in-plane stiffness of wood diaphragms is much lower. This factor must be taken into account when determining the response of wood buildings because diaphragm deformation alters how the loads are distributed to the vertical components of the LLRS below, as well as the entire performance of the building when subjected to lateral loads. For analysis purposes, engineers usually try to simplify the behaviour of the diaphragm into two categories: rigid diaphragms and flexible diaphragms. The standard assumption is that flexible diaphragms distribute the lateral loads to the LLRS underneath in proportion to tributary areas, whereas rigid diaphragms distribute loads in proportion to the stiffness of the LLRS elements. While one does not need to consider torsional effects when analysing flexible diaphragms, it is necessary to take such effects into account for rigid diaphragms due to the eccentricity between the centre of stiffness and the centre of mass. It should be noted that the NBC (National Research Council of Canada [NRC], 2020) requires designs to consider accidental torsion even for flexible diaphragms.

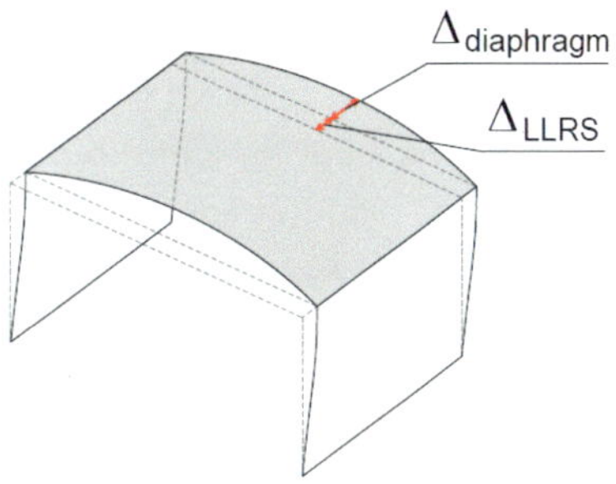

Figure 91. Deflection in the middle of the diaphragms, and that of the LLRS underneath (Moroder & Chen, 2022).

Currently, there are no criteria in the NBC (NRC, 2020) or in CSA O86 (CSA, 2019) for classifying diaphragms as rigid or flexible. Since the force distribution in a diaphragm and LLRS depends on the diaphragm flexibility, one must assess both stiffnesses to determine the flexibility classification of the diaphragm (Figure 91). When a diaphragm is flexible or rigid has been a point of discussion for a long time, and there is no consensus. Some building codes provide prescriptive detailing rules for rigid diaphragms. For example, in the U.S., ASCE7-22 (ASCE, 2022) assumes that a diaphragm is flexible when the displacement in the middle is at least twice that of the LLRS elements on both its side, i.e., $\Delta_{diaphragm} \geq 2\ \Delta_{LLRS}$. On the other hand, Eurocode 8, the seismic analysis and design code for most of Europe, classifies rigid diaphragms as those that have sufficiently large in-plane stiffness, as compared to that of the vertical LLRS, that the diaphragm has only a small effect on the distribution of the forces along the vertical structural elements (the LLRS). The flexibility and horizontal displacements of a rigid diaphragm should not exceed those resulting from the rigid diaphragm assumption by more than 10% anywhere along the diaphragm.

The 2020 NBC acknowledges the influence of the flexibility of the diaphragm on the overall response of the SFRS in single-storey building. The NBC requires that for single-storey buildings with steel deck or wood roof diaphragms designed with an R_d value greater than 1.5, and where the calculated maximum relative deflection of the diaphragm, Δ_D, under lateral loads exceeds 50% of the average storey drift of the adjoining vertical elements of the SFRS, Δ_{SFRS}, the design must account for the dynamic magnification of the inelastic response due to in-plane diaphragm deformations. In this case, one should design and detail the vertical elements of the SFRS for one of the following cases:

- Accommodate the anticipated magnified lateral deformations, taken as $R_oR_d\ (\Delta_{SFRS} + \Delta_D) - R_o\ \Delta_D$,
- Resist the forces magnified by $R_d\ (1 + \Delta_D/\Delta_{SFRS})/(R_d + \Delta_D/\Delta_{SFRS})$, or
- Conduct a special study.

In all cases, the roof diaphragm and chords should allow for in-plane shears and moments determined while taking into consideration the inelastic higher mode response of the structure.

The most common practice for the analysis and design of LWF diaphragms is to assume that they are flexible in their own plane and therefore distribute loads to the SFRS elements according to the tributary area. Another common practice is to design the diaphragm and the LLRS using the so-called envelope approach. This approach, implemented in some building codes (e.g., the 2021 International Building Code (IBC); International Code Council, 2021), requires completion of two analyses, one that assumes the diaphragm is flexible and another that assumes the diaphragm is rigid. The design of the main components of the diagrams and the SFRS subsequently makes use of the most critical results for each component from both analyses. This type of analysis is conservative and can overestimate the loads on some components of the diaphragm and the SFRS. As most mass timber diaphragms behave as semi-rigid components (Gagnon & Karacabeyli, 2019), mass timber buildings will benefit from an optimized analysis and diaphragm design. Such approaches, however, are complex and not yet codified, since the knowledge of mass timber diaphragm performance and design lags behind that of the other parts of the SFRS.

7.2 Capacity-Based Design Principles

The concept of capacity design is significant in the seismic design of timber buildings. This design approach is based on the simple understanding of how a structure sustains large deformations under severe earthquakes. By selecting certain modes of deformation for the SFRS, one can design and detail specific parts of it for yielding and energy dissipation during a seismic event under the imposed severe deformations. These critical regions of the SFRS, often termed "plastic hinges" or "dissipative zones", act as energy dissipaters to control the force levels in the structure. It is then possible to design all other structural elements as non-yielding with sufficient overstrength (capacity protected) greater than the probable strength of the plastic hinges. In other words, non-yielding elements, resisting actions originating in plastic hinges, must be designed for strength based on overstrength rather than on code-specified factored strength (resistance), which is used to determine the required strengths of the hinge regions. A typical explanation of the capacity-based design using a chain of ductile and capacity-protected elements appears in Figure 92.

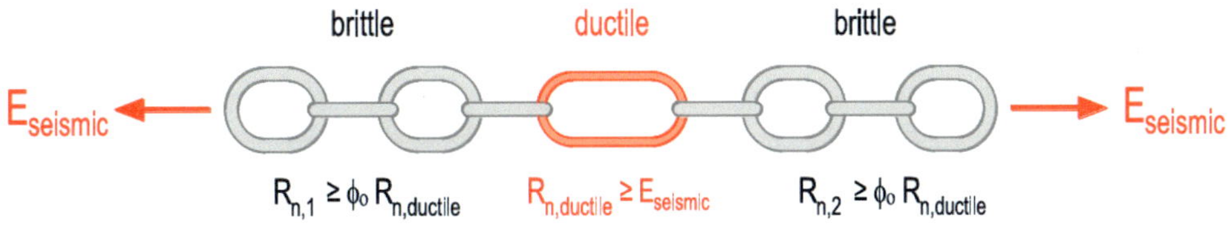

Figure 92. Capacity design and chain analogy ($E_{seismic}$ is the seismic demand, $R_{n,\ ductile}$ is the strength of the ductile element, ϕ_o is the overstrength factor of the ductile element, R_{n1},and R_{n2} are strengths of the brittle element) (Moroder, 2016).

Note that capacity design, as a design philosophy, enables the designer to "tell the structure what to do" and to desensitize it to the characteristics of an earthquake, which are, after all, unknown. Subsequent judicious detailing of all potential plastic regions will enable the structure to fulfill the designer's intentions. A capacity design approach is likely to ensure a more predictable and satisfactory inelastic response under variable earthquake conditions because a capacity-designed structure should not develop undesirable hinge mechanisms or modes of inelastic deformation that would result in collapse or prevent evacuation after a major seismic event. The approach is insensitive to earthquake characteristics, as far as the magnitude of inelastic deformations is concerned. When combined with appropriate detailing for ductility, capacity design will enable optimum energy dissipation via rationally selected plastic mechanisms.

In capacity-based design in steel structures, the design of the members typically allows them to yield before the connections; beam failure mechanisms are preferable since they can provide sufficient structural ductility without creating an undesirable mechanism of collapse. In timber structures, however, the failure of wood members in tension or bending is not favourable because of wood's brittle characteristics; consequently, all nonlinear deformations and energy dissipation in wood structures should occur in the connections. Thus, the steel and timber approaches are completely in opposition to one another.

The concept of capacity-based design has only recently been extended to wood-frame, heavy timber, and CLT structures (Popovski & Karacabeyli, 2008; Jorissen & Fragiacomo, 2011; Gavric et al., 2013; Shahnewaz et al., 2017; Casagrande et al., 2019). Also, it has only recently seen implementation in domestic and international wood design codes and standards. The next section presents design aspects in the codes and standards related to mass timber diaphragms.

7.3 Design Principles in Canadian, U.S., and International Codes and Standards

7.3.1 The Canadian Standard for Engineering Design in Wood

The 2019 edition of CSA O86, the Canadian Standard for Engineering Design in Wood (CSA, 2019), contains design provisions for CLT shear walls and diaphragms. It states that the lateral resistance of CLT diaphragms should be governed by the resistance of connections between the diaphragms and the supporting structure and by connections between the individual panels, calculated using methods of mechanics.

The CLT panels in the diaphragms should be capacity protected by designing them for the seismic forces that develop when energy-dissipative connections in shear walls reach the 95th percentile of their ultimate resistance. Similarly, the shear connections between the diaphragms and the walls beneath, as well as the connections between adjacent diaphragm panels, should be non-dissipative. Such connections should be capacity-protected by designing them to remain elastic under the force and displacement demands induced in them when the energy-dissipative connections in the system reach the 95th percentile of their ultimate resistance or target displacement. Alternatively, in both the CLT panels and the non-dissipative connections, one can determine the seismic design force by using a product of the force modification factors of $R_dR_o = 1.3$ as per the NBC (NRC, 2021).

The CSA O86 Standard also states that the diaphragm chords, struts, and other force-transfer elements, including those around openings, should be capacity protected. They should remain elastic under the force and displacement demands induced in them when the energy-dissipative connections between them and the SFRS reach the 95th percentile of their ultimate resistance or the design target displacement. Typical storey of a multi-storey CLT structure in platform construction with various connections between panels appears in Figure 93.

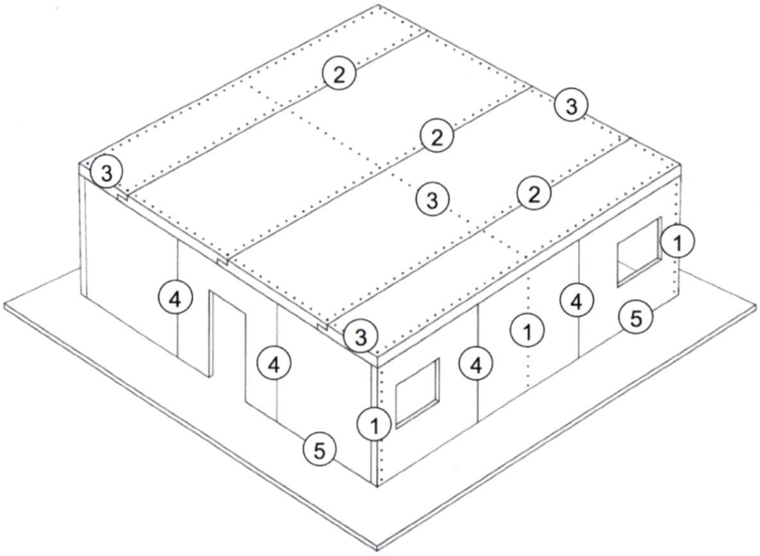

Figure 93. Typical storey of a multi-storey CLT structure with various connections between panels. Diaphragm connections 2 and 3 have to be elastic, while wall-to-wall connections 4 must be ductile.

Diaphragm chords, struts, and other force-transfer elements, including those around openings, should be capacity protected. They must remain elastic under the force and displacement demands induced in them when the energy-dissipative connections between them and the SFRS reach the 95th percentile of their ultimate resistance or the design target displacement.

7.3.2 Special Design Provisions for Wood and Seismic in the U.S.

The 2021 edition of the Special Design Provisions for Wood and Seismic (SDPWS) in the U.S. implemented design principles for mass timber diaphragms (ANSI/AWC, 2021). They require the estimation of diaphragm deflections using principles of engineering mechanics. The design of the CLT diaphragms should also proceed according to principles of engineering mechanics, using design values for wood members and connection as per U.S. National Design Specifications (ANSI/AWC, 2018). The diaphragm nominal shear capacity is defined as

4.5 times the 10-year load duration allowable stress design capacity of the dowel-type fastener used to transfer shear between CLT panels and between CLT panels and diaphragm boundary elements. To promote connection designs with fastener yielding modes and favourable overstrength, the calculated capacity of the dowel-type fastener in shear connections must also be controlled by yielding Mode IIIs or Mode IV when analyzed in accordance with the NDS (ANSI/AWC, 2018).

The allowable seismic and wind design shear capacities in the study were developed by reducing the nominal diaphragm shear capacity by factors of 2.8 and 2.0, respectively. These are the same adjustment factors as for nailed wood-frame diaphragms designed in accordance with SDPWS. These provisions also require connections that transfer diaphragm shear forces not be relied upon for transferring the diaphragm chord forces. According to capacity-based principles, the wood elements, steel elements, and chord splice connections must be designed for force equal to 2.0 times the diaphragm forces associated with the induced shear. An exception to this amplification is that dowel-type fasteners in chord splice connections can be designed for 1.5 times the forces associated with induced shear if they are designed to fail in yielding Mode IIIs or Mode IV. The force increase requirement, which needs added strength in diaphragm components other than the shear connections, aims to ensure that a minimum target overstrength of 2.8 times the allowable seismic design capacity in the diaphragm shear connections develops before other strength limit states. More information on the SDPWS provisions for diaphragm design is available in Breneman et al. (2022). For more comprehensive information about the new design procedures for CLT diaphragms in SDPWS, along with several solved design examples, see the newly released CLT Diaphragm Design Guide by Breneman et al. (2023), prepared by the Wood Works Wood Products Council in the U.S.

Figure 94. The test setup, with diaphragm specimen no. 2 ready for testing (Line et al., 2022).

The AWC conducted full-scale testing of two CLT diaphragms designed following the SDPWS provisions (Figure 94), reported in section 6.5.2 in this Guide (Line et al., 2022). The maximum loads achieved in the testing significantly exceeded the calculated nominal capacity of the diaphragms, mainly because alternative load paths developed in the diaphragms during the testing. The authors suggested refinements in the design procedure and detailing requirements to reduce this conservatism in the design.

7.3.3 European Codes and Standards

The current edition of Eurocode 8, the European Seismic Design Code (CEN, 2004b), contains only high-level capacity design procedures for timber structures. It recognizes three ductility classes among timber-based dissipative systems: DCL, DCM, and DCH, respectively referring to structures with low, medium, and high energy-dissipative structural behaviour. For the energy-dissipative connections in the structure, also called dissipative zones, Eurocode 8 requires that a dissipative zone shall be able to deform plastically for at least three fully reversed cycles at a static ductility ratio of 4 for ductility class medium (DCM) structures and at a static ductility ratio of 6 for ductility class high structures (DCH), without more than a 20% reduction of resistance. This is the first attempt of any code to link the local (connection) ductility to the global or system ductility.

A structural model may consider timber diaphragms rigid without further verification, if they can follow the code detailing rules for horizontal diaphragms and if their openings do not significantly affect the overall in-plane rigidity. The detailing rules do not allow the use of the factor of 1.2 for increasing the resistance of fasteners at the sheathing edges, which is otherwise allowed for some connections. When the sheets are staggered, it is also best to avoid using the increase factor of 1.5 for the nail spacing along the discontinuous panel edges. This provides a more conservative diaphragm design, which one could consider as capacity protection for the diaphragm, to some degree. To evaluate the distribution of shear forces in diaphragms, one should take into account the in-plane position of the vertical lateral load resisting elements. All sheathing edges that do not meet on framing members should be supported on and connected to a transverse blocking placed between the wooden beams. Blocking is also necessary in the horizontal diaphragms just above the lateral load resisting vertical elements (e.g., walls). It is important to ensure the continuity of beams, including the joists in areas where openings disturb the diaphragm. Without intermediate transverse blocking over the full height of the beams, the height-to-width ratio (h/b) of the timber beams should be less than four. In zones with a spectral acceleration higher than 0.2 g, the spacing of fasteners in areas of discontinuity should be reduced by 25%. Finally, when floors are considered rigid in plane, there should be no change of beams in the span direction over the supports (where horizontal forces are transferred to vertical elements e.g., shear-walls). Note that these provisions were intended mostly for light wood diaphragms, as mass timber diaphragms were in their infancy at the time.

A new edition of Eurocode 8 is expected in 2025. Its draft document (CEN, 2020) states that diaphragms and bracings in horizontal planes and quasi-horizontal planes (roofs) should be able to transmit, with appropriate overstrength when required, the effects of the design seismic action to the LLRSs to which they connect. Section 6.2.8(1) states that the design of the diaphragm may be considered satisfactory if, for the relevant resistance verifications, the seismic action effects in the diaphragm obtained from the analysis are multiplied by an overstrength factor. This overstrength factor depends on the failure mode of the diaphragm itself and the ductility class of the SFRS underneath. The ductility classes in this draft of the new edition of Eurocode 8 are labelled DC1, DC2, and DC3, for low, moderate, and highly ductile systems, respectively. If the failure mode of the diaphragm is a brittle one, then depending on the ductility class, the overstrength factor γ_d, provided in Table 6.1 of the draft document, should be taken as 1.0, 1.2, or 1.5 for systems in DC1, DC2, or DC3, respectively. For diaphragms with ductile failure mode, the overstrength factor should be taken as 1.0, 1.1, or 1.2 for systems in DC1, DC2, or DC3, respectively.

Designs can use glulam, solid wood panels, or LVL panels in shear walls, floor, and roof diaphragms, provided that they take measures to limit in-plane shrinkage in the direction perpendicular to the face grain. In floor and roof diaphragms, the designer should consider the different stiffness of glulam, solid wood panels, and LVL in both in-plane directions. CLT or glulam panels used in walls or diaphragms should be minimum 60 mm thick.

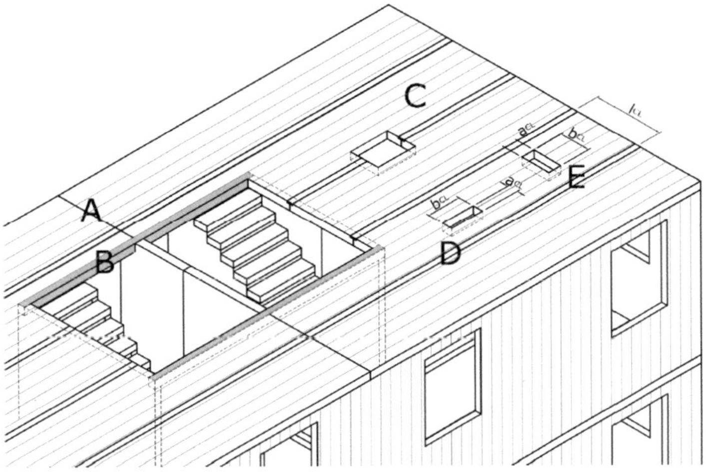

Figure 95. Openings in CLT diaphragms (CEN, 2020).

The draft document also has a new section 13.7 on CLT structures and a subsection 13.15.2 related to CLT floor and roof diaphragms. The panels in such diaphragms should be connected to each other and to the supporting walls or beams using metal fasteners such as screws or nails. Openings for stairs in horizontal diaphragms should be supported by walls or beams along the entire perimeter of the opening Figure 95 Detail A and B). It is important to check openings for installations placed across two CLT panels, like in Figure 95 (Detail C), to ensure the transmission of horizontal seismic forces in the panel-to-panel reduced length of joint. Holes pre-cut inside CLT panels for installations should have a longer side b_{CL} along the grain direction of the top CLT layer (Figure 95 Detail D), with the hole width $a_{CL} < 0.25\ l_{CL}$ (the width of the panel) and hole length $b_{CL} < 4\ a_{CL}$. It is best to avoid holes like the one in Figure 95 Detail E.

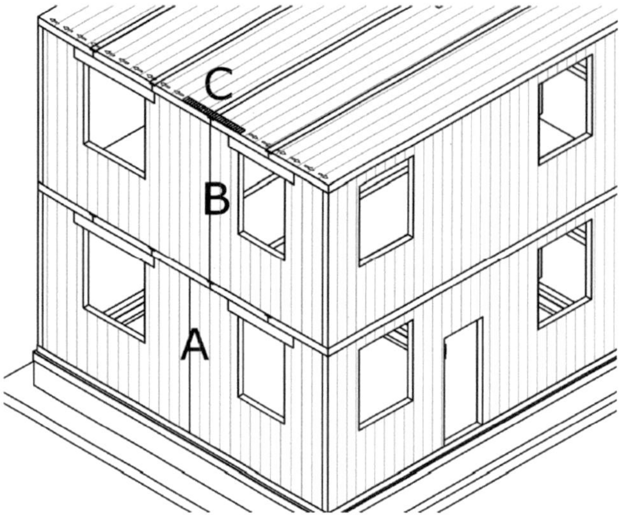

Figure 96. Location of horizontal joints in CLT diaphragms vs. vertical joints in CLT walls (CEN, 2020).

Joints between horizontal panels in CLT diaphragms should be staggered with respect to joints between vertical panels in CLT walls (Figure 96 detail A vs. B). Otherwise, additional measures, such as continuous beams or steel straps, are necessary to connect the floor panels above the wall and across the joint to transfer the tension forces due to the in-plane bending moments in the diaphragm (Figure 96 detail C).

Section 13.5(4) states that one can model timber-based floor and roof diaphragms supported by a primary timber structure as rigid in plane if their openings do not significantly affect the overall in-plane rigidity of the floor. A diaphragm with an aspect ratio of 2.0 or less and with openings of less than 10% of the floor area, not located along the perimeter, may be considered as rigid if it also satisfies one of the requirements below:

a) For all timber structural systems in Ductility Category 1 (DC1), the diaphragm and its connections should be designed to transfer the in-plane seismic shear to the primary structure using an overstrength factor γ_d of 1.5 instead of the values provided in Table 6.1 of that document.

b) For all structural types other than CLT and light-frame structures in DC2 and DC3 categories, the diaphragm and its connections should be designed to transfer the in-plane seismic shear to the primary structure using an overstrength factor 2.0 instead of the values for γ_d provided in Table 6.1 of that document.

c) For CLT and light-frame structures in DC2 and DC3 category, the diaphragm and its connections should be designed to transfer the in-plane seismic shear to the primary structure according to Formula below:

$$F_{Rd,b} \geq \frac{\gamma_{R,d}}{k_{deg}} \cdot \Omega_d \cdot F_{Ed,E} + F_{Ed,G} \tag{52}$$

where

$F_{Rd,b}$ is the design strength of the non-dissipative joint or structural element;

γ_{Rd} is the overstrength factor, equal to 1.6 except for high ductility moment-resisting frames with expanded tube fasteners and Densified Veneer Wood, for which it can be taken as 1.3;

k_{deg} is the strength reduction factor due to degradation under cyclic loading, taken as equal to 0.8;

$F_{Ed,E}$ is the action effect in the non-dissipative joint or member due to the design seismic action;

$F_{Ed,G}$ is the action effect in the non-dissipative joint or member due to the non-seismic actions in the design seismic situation; and

Ω_d is the minimum value of all overstrength ratios $\Omega_{d,i}$, calculated at each i_{th} storey. To calculate this factor, refer to formula (13.9) in CEN (2020).

One can consider TCC floor and roof diaphragms designed according to applicable European standards rigid in-plane if their openings do not significantly affect the overall in-plane rigidity of the floors. One can also assume that openings of less than 20% of the floor area do not significantly affect the overall in-plane rigidity. For the diaphragm to be considered rigid, the concrete topping should be at least 50 mm thick and should connect to all primary members.

8 CONCLUDING REMARKS

Diaphragms are fundamental parts of the lateral load resisting system (LLRS) of any building. They transfer horizontal loads to the LLRS and also tie all structural and non-structural elements together, providing integrity to the entire building. The current worldwide trend to build with mass timber has led to the introduction of this material in many diaphragm applications. Despite their fundamental role in building performance, there is thus far little research on the performance of mass timber diaphragms and their components under in-plane loads. In addition, design guidelines for mass timber diaphragms are not as comprehensive as those for other parts of mass timber buildings in the Canadian, U.S., and international codes and material standards. This document presents results from state-of-the-art experimental and analytical research related to mass timber diaphragms around the world. It also provides up-to-date information on methods of analysis for mass timber diaphragms and summarizes the available information on the design of mass timber diaphragms available in Canadian, U.S., and international codes and standards. This document should serve as a valuable reference for designers, architects, and other specifiers on the performance and design of mass timber diaphragms. For more comprehensive information about the design of CLT diaphragms using the latest U.S. special design provisions for wind and seismic (ANSI/AWC, 2021), along with design examples, please refer to the first edition of the *CLT Diaphragm Design Guide* (Breneman et al., 2023).

9 REFERENCES

American National Standards Institute & American Wood Council. (2018). *National design specification for wood construction*. American Forest and Paper Association. https://awc.org/publications/2018-nds/

American National Standards Institute & American Wood Council. (2021). *SDPWS: Special design provisions for wind and seismic*. American Forest and Paper Association. https://awc.org/publications/2021-sdpws/

American National Standards Institute & APA – The Engineered Wood Association. (2019). *Standard for performance-rated cross-laminated timber* (ANSI/APA Standard PRG 320). https://www.apawood.org/ansi-apa-prg-320

American Society of Civil Engineers. (2017). *Seismic rehabilitation and retrofit of existing buildings* (ASCE/SEI Standard 41-17). https://sp360.asce.org/PersonifyEbusiness/Merchandise/Product-Details/productId/233163464

American Society of Civil Engineers. (2022). *Minimum design loads and associated criteria for buildings and other structures* (ASCE/SEI Standard 7-22). https://www.asce.org/publications-and-news/asce-7

American Wood Council. (2015). *General dowel equations for calculating lateral connection values* (Technical Report 12). https://awc.org/wp-content/uploads/2021/12/AWC-TR12-1510.pdf

Ansys Inc. (2023). *A structural analysis & simulation software* [computer software].

Applied Technology Council. (1981). *ATC-7: Guidelines for the design of horizontal wood diaphragms.*

Ashtari, S. (2012). *In-plane stiffness of cross-laminated timber floors* [Master's Thesis, University of British Columbia]. UBC Theses and Dissertations. http://hdl.handle.net/2429/43504

ASTM International. (2019). *Standard test methods for cyclic (reversed) load test for shear resistance of vertical elements of the lateral force resisting systems for buildings* (ASTM Standard E2126). https://www.astm.org/e2126-19.html

ASTM International. (2021). *Standard test methods for single-bolt connections in wood and wood-based products* (ASTM Standard D5652). https://www.astm.org/d5652-21.html

ASTM International. (2022). *Standard test methods of static tests of lumber in structural sizes* (ASTM Standard D198). https://www.astm.org/d0198-22.html

ASTM International. (2023). *Standard test method for evaluating dowel-bearing strength of wood and wood-based products* (ASTM Standard D5764). https://www.astm.org/d5764-97ar18.html

Barbosa, A. R., Higgins, C., Sinha, A., Demeza, B. T., Rodrigues, L., Pei, S., van de Lindt, J., Blomberg, H., Zimmerman, R., McDonnell, E, Berman, J., Dolan, D., & Breneman, S. (2019). *Design and testing of shake-table specimen of cross-laminated timber and cross laminated timber-concrete composite diaphragms* (Report No. 18-01). Oregon State University.

Barbosa, A. R., Rodrigues, L. G., Sinha, A., Higgins, C., Zimmerman, R. B., Breneman, S., Pei, S., van de Lindt, J.W., Berman, J., & McDonnell, E. (2021). Shake-table experimental testing and performance of topped and untopped cross-laminated timber diaphragms. *Journal of Structural Engineering, 147*(4), Article 04021011. https://doi.org/10.1061/(ASCE)ST.1943-541X.0002914

Binational Softwood Lumber Council & Forest Innovation Investment. (2017). *Nail-laminated timber: Canadian design & construction guide.* https://www.naturallywood.com/wp-content/uploads/NLT_Canadian_Design_Construction_Guide.pdf

Brandner, R., Dietsch, P., Dröscher, J., Schulte-Wrede, M., Kreuzinger, H., & Sieder, M. (2017). Cross laminated timber (CLT) diaphragms under shear: Test configuration, properties and design. *Construction and Building Materials*, *147 (Aug)*, 312–327. https://doi.org/10.1016/j.conbuildmat.2017.04.153

Brandner, R., Flatscher, G., Ringhofer, A., Schickhofer, G., & Thiel, A. (2016). Cross laminated timber (CLT): Overview and development. *European Journal of Wood and Wood Products*, *74*(3), 331–351. https://doi.org/10.1007/s00107-015-0999-5.

Breneman S., McDonnell, E., Tremayne, B., Llanes, D., Houston, J., Gu, E., Zimmerman, R., & Montgomery, G. (2023). *CLT diaphragm design guide*. WoodWorks: Wood Products Council.

Breneman, S., McDonnell, E., & Zimmerman, R. B. (2016). *An approach to CLT diaphragm modeling for seismic design with application to a U.S. high-rise project* (WW-WSP-08). WoodWorks: Wood Products Council. https://www.woodworks.org/wp-content/uploads/Approach-to-CLT-Diaphragm-Modeling-for-Seismic-WoodWorks-Jan-2017.pdf

Breneman, S., McDonnell, E., Tremayne, B., Llanes, D., Houston, J., Gu, M., Zimmerman, R., Montgomery, G. (2022). *CLT diaphragm design for wind and seismic resistance using SDPWS 2021 and ASCE 7-22* (WW-WSP-25). WoodWorks: Wood Products Council. https://www.woodworks.org/wp-content/uploads/wood_solution_paper_clt_diaphragm.pdf

Bull, D. K. (2004). Understanding the complexities of designing diaphragms in buildings for earthquakes. *Bulletin of the New Zealand Society for Earthquake Engineering*, *37*(2), 70–88. https://doi.org/10.5459/bnzsee.37.2.70-88

Bull, D., & Henry, R. (2014). *Strut & tie* (Seminar Notes TR57). The New Zealand Concrete Society. https://cdn.ymaws.com/concretenz.org.nz/resource/resmgr/docs/nzcs/nzcs_tr57.pdf

Canadian Standards Association. (2019). *Engineering design in wood* (CSA Standard O86:19). https://www.csagroup.org/store/product/2702965/

Casagrande, D., Doudak, G., & Polastri, A. (2019). A proposal for the capacity-design at wall- and building-level in light-frame and cross-laminated timber buildings. *Bulletin of Earthquake Engineering 17*, 3139–3167. https://doi.org/10.1007/s10518-019-00578-4

Closen, M. (2017). *Performance of CLT connections under dynamic loading* [PDF slides]. MyTiCon Timber Connectors. https://mtcsolutions.com/wp-content/uploads/2019/04/Slides_-_Performance_of_CLT_Connections_under_Dynamic_Loading_-_USA.pdf

Computers and Structures Inc. (2016a). *SAP2000. Static and dynamic finite analysis of structures* [Computer software].

Computers and Structures Inc. (2016b). *ETABS. Extended 3D analysis of building system* [Computer software].

Consortium of Universities for Research in Earthquake Engineering. (2000). The CUREE-Caltech woodframe project [Scholarly project].

Cuerrier-Auclair, S. (2020). *Design guide for timber-concrete composite floors in Canada*. FPInnovations. https://web.fpinnovations.ca/tcc/

D'Arenzo, G., Casagrande, D., & Seim, W. (2021). Rigid or flexible? A numerical investigation on the in-plane behaviour of CLT floor diaphragms. World Conference on Timber Engineering (WCTE) 2021, Santiago, Chile. https://www.researchgate.net/publication/354339083_Rigid_or_flexible_A_numerical_investigation_on_the_in-plane_behaviour_of_CLT_floor_diaphragms

D'Arenzo, G., Casagrande, D., Reynolds, T., & Fossetti, M. (2019). In-plane elastic flexibility of cross laminated timber floor diaphragms. *Construction and Building Materials, 209*, 709–724. https://doi.org/10.1016/j.conbuildmat.2019.03.060

Dassault Systèmes. (2012). *Abaqus 6.12: Analysis user's manual: Volume II: Analysis*.

DeStefano, J. (2019). *Detailing cross laminated timber connections* (Timber Design Guide 2019-15). Timber Frame Engineering Council. https://www.tfguild.org/timber-frame-engineering-council/technical-bulletins/view/294/download

Deutsches Institut für Bautechnik. (2012). *Heco-Topix®-T and Heco-Topix®-CC screws for use in timber constructions* (European Technical Approval ETA-12/0132).

Deutsches Institut für Bautechnik. (2018). *Würth self-tapping screws* (European Technical Approval ETA-11/0190).

Diekmann, E. F. (1982). Design of wood diaphragms. *Journal of Materials Education, 8*(1–2).

Dlubal Software GmbH. (2016). *FEM structural analysis software* [Computer software].

Dolan, J. D., & Foschi, R. O. (1991). Structural analysis model for static loads on timber shear walls. *ASCE Journal of Structural Engineering, 117*(3), 851–861. https://doi.org/10.1061/(ASCE)0733-9445(1991)117:3(851)

Dubas, P. (1981) Einführung in die Norm 164 (1981) Holzbau [Introduction to Standard 164 (1981) Timber Structures] (Pub. Nr. 81–1). Institut für Baustatik und Stahlbau.

Earl, C. (2009). *Deflection of light frame wood diaphragms*. Washington State University, Pullman.

Ehlbeck, J. (1979). Load-carrying capacity and deformation characteristics of nailed joints. In *CIB-W18 Meeting Twelve* (Paper CIB-W18/12-7-1).

Ehlbeck, J., & Larsen, H. J. (1993). Eurocode 5 - Design of timber structures: Joints. In *Proceedings of the International Workshop on Wood Connectors* (pp. 9–23). Forest Products Society.

Ehlbeck, J., & Werner, H. (1988a). Design of joints with laterally loaded dowels. Proposals for improving the design rules in the CIB-Code and the draft Eurocode 5. In *CIB-W18 Meeting Twenty-One* (Paper CIB-W18/21-7-4).

Ehlbeck, J., & Werner, H. (1988b). Untersuchungen über die Tragfähigkeit von Stabdübelverbindungen [Investigations on the load-bearing capacity of rod dowel connections]. *European Journal of Wood and Wood Products, 46*(8), 281–288. https://doi.org/10.1007/BF02615055.

European Committee for Standardization. (1991). *Timber structures. Joints made with mechanical fasteners. General principles for the determination of strength and deformation characteristics* (CEN Standard EN 26891).

European Committee for Standardization. (2001). *Timber structures. Test methods. Cyclic testing of joints made with mechanical fasteners* (CEN Standard EN 12512).

European Committee for Standardization. (2004a). *Eurocode 5: Design of timber structures – Part 1-1: General – Common rules and rules for buildings* (CEN Standard EN 1995-1-1:2004).

European Committee for Standardization. (2004b). *Eurocode 8: Design of structures for earthquake resistance – Part 1: General rules, seismic actions and rules for buildings* (CEN Standard EN 1998-1:2004).

European Committee for Standardization. (2012). *Timber structures–Structural timber and glued laminated timber-Determination of some physical and mechanical properties* (CEN Standard EN 408).

European Committee for Standardization. (2020). *Eurocode 8: Design of structures for earthquake resistance; Part 1-2: Rules for new buildings* (CEN Standard prEN 1998-1-2:2020 (E), SC8.T2 Final Document).

Falk, R. H., & Itani, R. Y. (1998) Finite Element Modeling of Wood Diaphragms. *ASCE Journal of Structural Engineering 115 (3)* https://doi.org/10.1061/(ASCE)0733-9445(1989)115:3(543)

Fleischman, R. B., & Farrow, K. T. (2001). Dynamic behavior of perimeter lateral-system structures with flexible diaphragms. *Earthquake Engineering & Structural Dynamics, 30*(5), 745–763. https://doi.org/10.1002/eqe.36

Foliente, G. C. (1996). Issues in seismic performance testing and evaluation of timber structural systems. In *Proceedings of the International Wood Engineering Conference; October 28–31, 1996; New Orleans, USA*.

Follesa, M., Christovasilis, I. P., Vassallo, D., Fragiacomo, M., & Ceccotti, A. (2013). Seismic design of multi-storey cross laminated timber buildings according to Eurocode 8. *Ingegneria Sismica, 20*(4), 27–53.

Foschi, R. O., & Bonac, T. (1977). Load-slip characteristics for connections with common nails. *Journal of Wood Science, 9*(3), 118–123.

Gavric, I., Ceccotti, A., & Fragiacomo, M. (2015). Cyclic behavior of typical screwed connections for cross-laminated (CLT) structures. *European Journal of Wood and Wood Products, 73*(2), 179–191. http://dx.doi.org/10.1007/s00107-014-0877-6

Gavric, I., Fragiacomo, M., & Ceccotti, A. (2013). Capacity seismic design of X-LAM wall systems based on connection mechanical properties. In R. Görlacher (Ed.), *CIB - W18 Meeting Proceedings, Meeting 46, 26–29 August 2013, Vancouver, Canada* (Paper CIB-W18/46-15-2, pp. 285–298).

Ghosh, S. K. (2016, March 1). Alternative diaphragm seismic design force level of ASCE 7-16. *Structure Magazine*, 18–23.

Hanna, D., & Tannert, T. (2021). Glulam connections assembled with screws in different installation angles. *Maderas. Ciencia y tecnología, 23*. https://doi.org/10.4067/s0718-221x2021000100454

Heine, C. P., & Dolan, J. D. (2001). A new model to predict the load-slip relationship of bolted connections in timber. *Wood and Fiber Science, 33*(4), 534–549.

Hossain, A., Danzig, I., & Tannert, T. (2016). Cross-laminated timber shear connections with double-angled self-tapping screw assemblies. *ASCE Journal of Structural Engineering, 142*(11). https://doi.org/10.1061/(ASCE)ST.1943-541X.0001572

Hossain, A., Popovski, M., & Tannert, T. (2018). Cross-laminated timber connections assembled with a combination of screws in withdrawal and screws in shear. *Engineering Structures, 168*, 1–11. https://doi.org/10.1016/j.engstruct.2018.04.052

Hossain, A., Popovski, M., & Tannert, T. (2019). Group effects for shear connections with self-tapping screws in CLT. *Journal of Structural Engineering, 145*(8), 1–9. https://doi.org/10.1061/(ASCE)ST.1943-541X.0002357

International Code Council. (2021). *International Building Code IBC 2021*.

International Standards Organization. (2003). *Timber structures — Joints made with mechanical fasteners — Quasi-static reversed-cyclic test method* (ISO 16670:2003). https://www.iso.org/obp/ui/#iso:std:iso:16670:ed-1:v1:en

Jalilifar, E., Koliou, M., & Pang, W. (2021). Experimental and numerical characterization of monotonic and cyclic performance of cross-laminated timber dowel-type connections. *Journal of Structural Engineering, 147(7)*, Article 04021102. https://doi.org/10.1061/(ASCE)ST.1943-541X.0003059

Jockwer R., & Jorissen A. (2018). Stiffness and deformation of connections with dowel-type fasteners. In C. Sandhaas, J. Munch-Andersen, & P. Dietsch (Eds.), *Design of Connections in Timber Structures: A state-of-the-art report by COST Action FP1402 / WG3* (pp. 95–126), Shaker Verlag Aachen.

Jockwer, R., Caprio, D., & Jorissen, A. (2021). Evaluation of parameters influencing the load-deformation behaviour of connections with laterally loaded dowel-type fasteners, *Wood Material Science & Engineering, 17*(1), 6–19. https://doi.org/10.1080/17480272.2021.1955297

Jorissen, A., & Fragiacomo, M. (2011). General notes on ductility in timber structures. *Engineering Structures Volume, 33*(11), 2987–2997. https://doi.org/10.1016/j.engstruct.2011.07.024

Kamiya, F. (1990). Horizontal plywood sheathed diaphragms with openings: Static loading tests and analysis. In H. Sugiyama (Ed.), *Proceedings of the 1990 International Timber Engineering Conference, Tokyo, Japan*.

Kamiya, F., & Itani, R.Y. (1998). Design of wood diaphragms with openings. *Journal of Structural Engineering, 124*(7): 839–848. https://doi.org/10.1061/(ASCE)0733-9445(1998)124:7(839)

Karacabeyli, E., & Ceccotti, A. (1996). Quasi-static reversed-cyclic testing of nailed joints. In R. Görlacher (compiler), *Proceedings of the CIB-W18 Meeting Twenty-Nine* (Paper CIB-W18/29-7-7).

Karacabeyli, E., & Gagnon, S. (2019). *Canadian CLT Handbook* (2nd ed.) (Special Publication SP-532E). FPInnovations.

Kessel, M.H., & Schönhoff, T. 2001. *Entwicklung eines Nachweisverfahrens für Scheiben auf der Grundlage von Eurocode 5 und DIN 1052 neu* [Development of a design methodology for diaphragms based on Eurocode 5 and DIN1052:2008]. Fraunhofer IRB Verlag.

Kim, M. T. (2021, April 28). Strut-and-tie model: Part 1 – Basics. *Midas Bridge*. https://www.midasbridge.com/en/blog/bridgeinsight/strut-and-tie-model-part-1-basics.

Kode, A., Amini, M. O., van de Lindt, J. W., & Line, P. (2021). Lateral load testing of a full-scale cross-laminated timber diaphragm. *Practice Periodical on Structural Design and Construction, 26*(2), Article 04021001. https://doi.org/10.1061/(ASCE)SC.1943-5576.0000566

Krenn, H., & Schickhofer, G. (2009). Joints with inclined screws and steel plates as outer members. In R. Görlacher (compiler), *CIB-W18, Meeting Forty-Two, Dübendorf, Switzerland* (Paper CIB-W18/42-7-2).

Lawson, J., Breneman, S., & Lo Ricco, M. (2023). Wood diaphragm deflections. I: Generalizing standard equations using mechanics-based derivations for panel construction. *Journal of Architectural Engineering, 29*(3), Article 04023019. https://doi.org/10.1061/JAEIED.AEENG-1573

Lemaître, R., Bocquet, J.-F., Schweigler, M., & Bader, T. K. (2018, August 13–16). Beam-on-foundation (BOF) modelling as an alternative design method for timber joints with dowel-type fasteners – Part 1: Strength and stiffness per shear plane of single-fastener joints. In R. Görlacher (Ed.), *INTER: International Network on Timber Engineering Research, Proceedings, Meeting 51* (pp. 241–255, Paper 51-7-13).

Line, P., Nyseth, S., & Waltz, N. (2021). Full-scale cross-laminated timber diaphragm evaluation. I: Design and full-scale diaphragm testing. *Journal of Structural Engineering, 148*(5), Article 04022037. https://doi.org/10.1061/(ASCE)ST.1943-541X.0003308

Loss, C., & Frangi, A. (2017). Experimental investigation on in-plane stiffness and strength of innovative steel-timber hybrid floor diaphragms. *Engineering Structures, 138,* 229–244. https://doi.org/10.1016/j.engstruct.2017.02.032

Loss, C., Rossi, S., & Tannert, T. (2018). In-plane stiffness of hybrid steel–cross-laminated timber floor diaphragms. *Journal of Structural Engineering, 144*(8), Article 04018128. https://doi.org/10.1061/(ASCE)ST.1943-541X.0002105

Malone, R. T., & Rice, R. W. (2012). *The analysis of irregular shaped structures: Diaphragms and shear walls.* McGraw Hill.

Masse, D. I., Salinas, J. J., & Turnbull, J. (1989). *Lateral strength and stiffness of single and multiple bolts in glued-laminated timber loaded parallel to grain* (Unpublished contract No. C-029). Engineering and Statistics Research Centre, Research Branch, Agriculture Canada.

McKenna, F., Fenves, G. L., Scott, M. H., & Jeremic, B. (2000). *Open system for earthquake engineering simulation* (OpenSees). Berkeley, USA.

Moehle, J. P., Hooper, J. D., Kelly, D. J., & Meyer, T. R. (2010). *Seismic design of cast-in-place concrete diaphragms, chords, and collectors: A guide for practicing engineers* (NIST GCR 10-917-4). National Institute of Standards and Technology, U.S. Department of Commerce.

Moroder, D. (2016). *Floor diaphragms in multi-storey timber buildings* [Doctoral dissertation, University of Canterbury, New Zealand]. UC Research Repository. http://hdl.handle.net/10092/11893

Moroder, D., & Chen, Z. (2022). Chapter 6.2. Diaphragms. In Z. Chen, D. Tung, & E. Karacabeyli (Eds.), *Modelling guide for timber structures* (Special Publication SP-544). FPInnovations.

Moroder, D., Smith, T., Pampanin, S., Palermo, A., & Buchanan, A. H. (2014). *Design of floor diaphragms in multi-storey timber buildings* [Conference presentation]. International Network on Timber Engineering Research 2014, Bath, England.

Moroder, D., Smith, T., Pampanin, S., Palermo, A., & Buchanan, A. H. (2015). *Design of floor diaphragms in multi-storey timber buildings* [Conference presentation]. New Zealand Society of Earthquake Engineering Conference 2015, Rotorua, New Zealand.

Muñoz, W., Mohammad, M., Salenikovich, A., & Quenneville, P. (2008). Need for a harmonized approach for calculations of ductility of timber assemblies. In R. Görlacher (compiler), *CIB-W18, Meeting Forty-One, St. Andrews, Canada* (Paper CIB-W18/41-15-1).

Nakaki, S.D. (2000). *Design guidelines for precast and cast-in-place concrete diaphragms* (NEHRP Professional Fellowship Report). Earthquake Engineering Research Institute.

National Research Council of Canada. (2021). *National Building Code of Canada 2020*. Canadian Commission on Building and Fire Code.

Pei, S., van de Lindt, J. W., Barbosa, A. R., Berman. J. W., McDonnell, E., Dolan, J. D., Blomgren, H.-E., Zimmerman, R. B., Huang, D., & Wichman, S. (2019). Experimental seismic response of a resilient 2-story mass-timber building with post-tensioned rocking walls. *Journal of Structural Engineering, 145*(11), Article 04019120. https://doi.org/10.1061/(ASCE)ST.1943-541X.0002382

Popovski, M., & Karacabeyli, E. (2008, October 12–17). *Force modification factors and capacity design procedures for braced timber frames* [Conference presentation]. 14th World Conference on Earthquake Engineering, Beijing, China.

Popovski, M., Tung, D., & Chen, Z. (2022). Section 5.3. Structural analysis and design. In E. Karacabeyli & C. Lum (Eds.), *Technical Guide for Design and Construction of Tall Wood Buildings in Canada* (Special Publication SP-543E). FPInnovations.

Ringhofer, A., Brandner, R., & Blaß, H. J. (2018). Cross laminated timber (CLT): Design approaches for dowel-type fasteners and connections. *Engineering Structures, 171*, 849–861. https://doi.org/10.1016/j.engstruct.2018.05.032

Scarry, J. M. (2014). *Floor diaphragms – Seismic bulwark or Achilles' heel* [Conference presentation]. New Zealand Society for Earthquake Engineering Conference 2014, Wellington, New Zealand.

Schlaich, J., Schafer, K., & Jennewein, M. (1987). Toward a consistent design of structural concrete. *PCI Journal, 32*(3), 74–150. https://doi.org/10.15554/pcij.05011987.74.150

Shahnewaz, Md., Tannert, T., Alam, M. S., & Popovski, M. (2017, April 6–8). *Capacity-based design for cross-laminated timber buildings* [Conference presentation]. Structures Congress 2017, Denver, Colorado. http://dx.doi.org/10.1061/9780784480427.034

Smith, P. C., Dowrick, D. J., & Dean, J. A. (1986). Horizontal timber diaphragms for wind and earthquake resistance. *Bulletin of the New Zealand Society for Earthquake Engineering, 19*(2): 135–142. https://doi.org/10.5459/bnzsee.19.2.135-142

Spickler, K., Closen, M., Line, P., & Pohil, M. (2015). *Cross laminated timber: Horizontal diaphragm design example* [White paper]. Structurlam, MyTiCon Timber Connectors, & American Wood Council. https://mtcsolutions.com/wp-content/uploads/2019/04/CLT_Horizontal_Diaphragm_Design_Example.pdf

Standards New Zealand. (2004). *Structural design actions Part 5: Earthquake actions - New Zealand* (NZS Standard 1170.5).

Standards New Zealand. (2022). *Timber structures* (NZS AS Standard 1720:1:2022).

Sullivan, K. (2017). *Behavior of cross-laminated timber diaphragm panel-to-panel connections with self-tapping screws* [Master's thesis, Oregon State University]. ScholarsArchive@OSU. https://ir.library.oregonstate.edu/concern/graduate_thesis_or_dissertations/n009w540c

Sullivan, K., Miller, T. H., & Gupta, R. (2018). Behavior of cross-laminated timber diaphragm connections with self-tapping screws. *Engineering Structures, 168*, 505–24. https://doi.org/10.1016/j.engstruct.2018.04.094

Swiss Society of Engineers and Architects. (2012). *Timber Structures* (SIA Standard 265).

Taylor, B. (2019). *In-plane shear performance of cross-laminated timber and cross-laminated timber concrete composite diaphragm connection systems* [Master's thesis, Oregon State University]. ScholarsArchive@OSU. https://ir.library.oregonstate.edu/concern/graduate_thesis_or_dissertations/bn999d16r

Taylor, B., Barbosa, A. R., & Sinha, A. (2020). Cyclic performance of in-plane shear cross-laminated timber panel-to-panel surface spline connections. *Engineering Structures, 218*, Article 110726. https://doi.org/10.1016/j.engstruct.2020.110726

Uibel T., & Blaß, H. J. (2007). Edge joints with dowel type fasteners in cross laminated timber. In R. Görlacher (compiler), *CIB-W18 Meeting Forty, Bled, Slovenia* (Paper CIB-W18/40-7-2).

Uibel, T., & Blaß, H. J. (2006). Load carrying capacity of joints with dowel type fasteners in solid wood panels. In R. Görlacher (compiler), *CIB-W18 Meeting Thirty-Nine, Florence, Italy* (Paper CIB-W18/39-7-5).

van de Lindt, J. W., Furley, J., Amini, M. O., Pei, S., Tamagnone, G., Barbosa, A. R., Rammer, D., Line, P., Fragiacomo, M., & Popovski, M. (2019). Experimental seismic behavior of a two-story CLT platform building. *Engineering Structures*, 183(15), 408–22. https://doi.org/10.1016/j.engstruct.2018.12.079

Wallner-Novak, M., Koppelhuber, J., & Pock, K. (2014). *Cross-laminated timber structural design: Basic design and engineering principles according to Eurocode*. ProHolz Austria.

Yasumura, M., & Kawai, N. (1998). Estimating seismic performance of wood-framed structures. In J. Natterer & J.-L. Sandoz (Eds.), *Proceedings of the 5th World Conference on Timber Engineering (WCTE 1998)*, vol. 2 (pp. 564–71).

Yasumura, M., Murota, T., & Sakai, H. (1987). Ultimate properties of bolted joints in glued-laminated timber. In *CIB-W18A, Meeting Twenty, Dublin, Ireland* (Paper CIB-W18/20-7-3).

Zahn, J. J. (1991). Design equation for multiple-fastener wood connections. *Journal of Structural Engineering, 117*(11), 3477–3486. https://doi.org/10.1061/(ASCE)0733-9445(1991)117:11(3477)

Made in the USA
Las Vegas, NV
03 November 2026

57424980R00076